Recent Modelling Approaches in Applied Energy Economics

INTERNATIONAL STUDIES IN ECONOMIC MODELLING

Series Editor
Homa Motamen–Scobi

Executive Director
Canadian Imperial Bank of Commerce
Securities Europe Ltd
London

Modelling in the OECD Countries
H. Motamen
Modelling the Labour Market
M. Beenstock
Input-Output Analysis
M. Ciaschini
**Modelling of Disequilibrium and Shortage
in Centrally Planned Economies**
C. Davis and W. Charemza

Recent Modelling Approaches in Applied Energy Economics

Edited by
Olav Bjerkholt
Øystein Olsen
and
Jon Vislie

CHAPMAN AND HALL
LONDON • NEW YORK • TOKYO • MELBOURNE • MADRAS

UK	Chapman and Hall, 2–6 Boundary Row, London SE1 8HN
USA	Chapman and Hall, 29 West 35th Street, New York NY10001
JAPAN	Chapman and Hall Japan, Thomson Publishing Japan, Hirakawacho Nemoto Building, 7F, 1-7-11 Hirakawa-cho, Chiyoda-ku, Tokyo 102
AUSTRALIA	Chapman and Hall Australia, Thomas Nelson Australia, 480 La Trobe Street, PO Box 4725, Melbourne 3000
INDIA	Chapman and Hall India, R. Seshadri, 32 Second Main Road, CIT East, Madras 600 035

First edition 1990

© 1990 Chapman and Hall Ltd

Typeset in 10/12 Sabon by
KEYTEC, Bridport, Dorset
Printed in Great Britain by
St Edmundsbury Press, Bury St. Edmunds, Suffolk

ISBN 0 412 353407

All rights reserved. No part of this publication may be reproduced or transmitted, in any form or by any means, electronic, mechanical, photocopying, recording or otherwise, or stored in any retrieval system of any nature, without the written permission of the copyright holder and the publisher, application for which shall be made to the publisher.

British Library Cataloguing in Publication Data
Recent modelling approaches in applied energy economics.—
(International studies in economic modelling)
1. Energy approaches. Models
I. Bjerkholt, Olav. II. Olsen, Øystein. III. Vislie, Jon.
333. 79011

ISBN 0-412-35340-7

Library of Congress Cataloging-in-Publication Data
available

Contents

Contributors vii
Acknowledgements ix
Introduction to the series xi
Preface xiii

PART ONE The European Gas Market

1 The Western European gas market: deregulation and supply competition 3
Olav Bjerkholt, Eystein Gjelsvik and Øystein Olsen

2 Residential energy demand – the evolution and future potential of natural gas in Western Europe 29
Sarita Bartlett, Steinar Strøm and Øystein Olsen

3 The European gas market as a bargaining game 49
Michael Hoel, Bjart Holtsmark and Jon Vislie

4 Bargaining, vertical control, and (de)regulation in the European gas market 67
Jon Vislie

5 Environmental effects of a transition from oil and coal to natural gas in Europe 87
Knut H. Alfsen, Lorents Lorentsen and Karine Nyborg

PART TWO Management of National Petroleum Resources

6 Petroleum resources and the management of national wealth 103
Iulie Aslaksen, Kjell Arne Brekke, Tor Arnt Johnsen and Asbjørn Aaheim

7	Oil and gas revenues and the Norwegian economy in retrospect: alternative macroeconomic policies Ådne Cappelen and Eystein Gjelsvik	125
8	The resource rent for Norwegian natural gas Rolf Golombek and Michael Hoel	153
9	Social discount rates for Norwegian oil projects under uncertainty Diderik Lund	171
10	The choice between hydro and thermal power generation under uncertainty T. Ø. Kobila	187
11	The management of jointly produced exhaustible resources Jon Vislie	207

PART THREE The World Oil Market and Macroeconomic Performance

12	The options for independent oil-exporting countries in the 1990s Kjell Berger, Olav Bjerkholt and Øystein Olsen	221
13	Business cycles and oil price fluctuations: some evidence for six OECD countries Knut Anton Mork, Hans Terje Mysen and Øystein Olsen	239
	Index	261

Contributors

Asbjørn Aaheim	Central Bureau of Statistics, PO Box 8131 Dep, 0033 Oslo, Norway
Knut H. Alfsen	Central Bureau of Statistics, Oslo, Norway
Iulie Aslaksen	Central Bureau of Statistics, Oslo, Norway
Sarita Bartlett	Lawrence Berkeley Laboratory, University of California, CA94720, USA
Kjell Berger	Central Bureau of Statistics, Oslo, Norway
Olav Bjerkholt	Central Bureau of Statistics, Oslo, Norway
Kjell Arne Brekke	Central Bureau of Statistics, Oslo, Norway
Ådne Cappelen	Central Bureau of Statistics, Oslo, Norway
Eystein Gjelsvik	Central Bureau of Statistics, Oslo, Norway
Rolf Golombek	Centre for Applied Research, Department of Economics, University of Oslo, PO Box 1095 Blindern, 0317 Oslo, Norway
Michael Hoel	Centre for Applied Research, Department of Economics, University of Oslo, Norway
Bjart Holtsmark	Central Bureau of Statistics, Oslo, Norway
Tor Arnt Johnsen	Central Bureau of Statistics, Oslo, Norway
Tom Lindstrøm	Department of Mathematics, University of Oslo, PO Box 1053, 0316 Oslo, Norway

	Contributors
Lorents Lorentsen	Central Bureau of Statistics, Oslo, Norway
Diderik Lund	Centre for Applied Research, Department of Economics, University of Oslo, Norway
Knut Anton Mork	Owen Graduate School of Management, Vanderbilt University, Nashville, Tennessee, USA
Hans Terje Mysen	Central Bureau of Statistics, Oslo, Norway
Karine Nyborg	Central Bureau of Statistics, Oslo, Norway
Øystein Olsen	Central Bureau of Statistics, Oslo, Norway
Bernt Øksendal	Department of Mathematics, University of Oslo, Norway
Steinar Strøm	Centre for Applied Research, Department of Economics, University of Oslo, Norway
Jon Vislie	Norwegian School of Management, PO Box 580, 1301 Sandvika, Norway

Acknowledgements

The content of this book is the outcome of several years research in petroleum economics by the Central Bureau of Statistics of Norway and the Centre for Applied Research, University of Oslo. The research activities have had a common funding in the Oil and Society programme of the Norwegian Research Council for Applied Social Science, which is gratefully acknowledged. Most of the papers were presented at a joint workshop in Åsgårdstrand in June 1989 under the auspices of the Oil and Society programme and the editors wish to thank David O. Wood, Center for Energy Policy Research, MIT; Knut Anton Mork, Owen Graduate School, Vanderbilt University; and David Newbery, Department of Applied Economics, University of Cambridge for their very valuable advice and criticism on this occasion.

Introduction to the series

There has been a growing dependence in the past two decades on modelling as a tool for better understanding of the behaviour of economic systems, and as an aid in policy and decision making. Given the current state of the art globally, the introduction of a series such as this can be seen as a timely development. This series will provide a forum for volumes on both the theoretical and applied aspects of the subject.

International Studies in Economic Modelling is designed to present comprehensive volumes on modelling work in various areas of the economic discipline. In this respect one of the fundamental objectives is to provide a medium for ongoing review of the progression of the field.

There is no doubt that economic modelling will figure prominently in the affairs of government and in the running of the private sector, in efforts to achieve a more rational and efficient handling of economic affairs. By formally structuring an economic system, it is possible to simulate and investigate the effect of changes on the system. This in turn leads to a growing appreciation of the relevance of modelling techniques. Our aim is to provide sufficient space for authors to write authoritative handbooks, giving basic facts with an overview of the current economic models in specific areas and publish a useful series which will be consulted and used as an accessible source of reference.

The question may arise in some readers' minds as to the role of this series *vis-à-vis* other existing publications. At present, no other book series possesses the characteristics of *International Studies in Economic Modelling* and as such cannot fill the gap that will be bridged by it. Those journals which focus on this area do not present an exhaustive and comprehensive overview of a particular subject and all the developments in the field. Other journals which may contain economic modelling papers are not sufficiently broad to publish volumes on all aspects of modelling in a specific area, which this series is designed to cover.

A variety of topics will be included encompassing areas of both micro- and macroeconomics, as well as the methodological aspects of model construction. Naturally, we are open to suggestions from all readers of, and contributors to, the series regarding its approach and content.

Finally, I would like to thank all those who have helped the launch of this series. The encouraging response received from authors who have contributed the forthcoming volumes and from the subscribers to the series has indicated the need for such a publication.

Homa Motamen–Scobi
London
December 1987

Preface

In 1990 both OPEC and the OECD will celebrate their thirtieth anniversaries. OPEC was founded – rather unnoticed – by oil-producing countries still struggling to gain control over national petroleum resources. Future members were still under colonial rule. The foremost aim of the new organization – years before it was able to make metropolitan newspaper headlines – was stabilizing oil prices. Stability in those days meant preventing oil prices from falling in real terms. The OECD was formed by mostly mature industrial economies marking the normalization of the postwar international economy after years of reconstruction, strict trade regulations, etc. The aim of the new organization was to promote 'the highest sustainable growth and employment' in member countries. Incidentally, 1960 was also the year which gave birth to a more loosely defined block in the world community, namely the underdeveloped countries, as the African colonial empires finally broke up.

The two organizations became adversaries in the 1970s in the power struggle over the energy flows of the world. It should not be forgotten, perhaps, that the first period in the life of the two organizations, i.e. from 1960 to 1973, must be counted as a success both with regard to energy price stability and growth of industrialized countries. After the dramatic events in the oil market in 1973 and the ensuing price increases, OPEC appeared to be much less concerned with stability of oil prices than with how high the oil rent could become! What goes up, however, has to come down. The OECD, on the other hand, seemed in a strange parallel movement to lose interest in promoting the highest possible employment and growth. Unemployment, especially in Europe, rose to unprecedented heights, while the energy price shock of OPEC I figured as the cause that forced the world economy off-track. The developing countries which started out on the difficult path towards industrialization, agricultural modernization and increased standards of living – in short, development – became the innocent victims of the shift away from the stability and growth of the 1960s.

The 1980s have brought lessons – for everyone. The decade started out with the second oil price shock. It has often been described as a unilateral action by OPEC. It is closer to the truth to depict OPEC II, as it was called, as a chain of events where actions were taken by a number of actors on assumptions that apparently were widely shared, but still were, above all, far from the truth. The false assumptions took the form for instance of mainstream long-term oil price forecasts, massively above what seemed likely only a few years later. OPEC's supreme control of the world oil market did not last long after the high-riding in 1979–80. The common belief in the 1970s of the coming crisis of exhaustion of the world's petroleum resources quickly gave way when the greater industrial powers managed to reduce their oil consumption while national oil production and independent oil exporters stole market shares away from OPEC. A remarkable fact, however, about OPEC is the cohesion among its members, in spite of war, antagonism, heterogenous economic interests and other less than negligible differences. OPEC has survived to disprove all rumours of its immediate demise.

The 1980s have also firmly established a major role for natural gas in the world's energy supply. How great this role might become will perhaps be even more evident at the end of the 1990s. The Western European gas market is still at an early stage of development with regard to infrastructure both of a technical kind and with regard to the market. A major event on the way is the forthcoming deregulation within the EC, and supposedly also including its energy markets. The increasing role of natural gas in the global energy supply is a result not only of the fact that there turned out to be much greater supplies than earlier assumed within affordable reach of the more concentrated urban and industrial areas of the world, but also from the concern with environmental dangers. While the 1970s were concerned with the coming exhaustion of limited resources such as petroleum, the environmental focus has shifted towards considering the suffocation, poisoning, spreading of cancer and other ways of dying as an indirect result of human civilization as the limit to sustainability. By marginal comparison this shift has benefited natural gas as an energy source above oil and coal. The last decade of the twentieth century will be entered, however, with no agreed solutions or effective policies designed to counteract the global threats such as the greenhouse effect and ozone layer destruction.

Many of the OECD countries, and the larger European countries in particular, pursued policies in the early 1980s that needlessly prolonged the austerity ventured upon as a cure-all in the mid-1970s. Unemployment turned out to be much harder to bring down than to increase, and the reduction of energy prices would not by itself do much for employment. While OPEC's power has been reduced as the organization has had to yield to market forces, the OECD has also over time become less of a decision maker in international economic policy. The economic summits comprising only a select few of the OECD countries have taken over the policy-coordinating role. After 30 years both OPEC and the OECD have experienced a full cycle of ups and downs with regard to the respective aims of the organizations.

This book is written from a Norwegian viewpoint. Norway, as one of the smaller OECD countries, fell into a bonanza of petroleum wealth from the vast territory adjoined to Norway after the dividing up of the continental shelf in the 1960s. The first quantities of oil from the Norwegian continental shelf were extracted in 1971, while gas production started in 1975. This 'timing' of production meant that Norway gained significantly from the increasing energy prices during the 1970s. While most other countries were hit hard by increasing bills on oil imports, Norway was able to reap large incomes from its petroleum exports. This was an important factor underlying the 'counter-cycle policy' that was pursued by the Norwegian government in the second half of the 1970s, keeping up economic growth and preventing the rise in employment that occurred in many other OECD countries. Even with the more sluggish oil market in the early 1980s the expectations held by most economists and planners in Norway in this period was that of continued increase in incomes from petroleum production.

That perception changed in an abrupt way in the winter of 1985–86 when the oil price collapsed and after some months 'stabilized' at $US15–18 per barrel. For an oil-exporting country like Norway the impact from the price fall was quite dramatic. The economic rent from petroleum production was brought close to zero. Furthermore, the events brought to the surface clear signs of 'Dutch disease' in the Norwegian economy: the activities in the North Sea and the spending of oil incomes have crowded out other onshore industries. To reverse the changes in industrial structure when oil incomes fail may be a very painful process.

So, in the late 1980s the situation both in the energy markets in general and for Norway as an oil producer is very different from the outlook ten years earlier. Also, the focus in economic thinking and analysis has changed considerably. There is no longer much concern about energy reserves representing 'limits to growth', and the market power of OPEC is not as dominant as in the 1970s. More significant, however, are the shifting trends and strong fluctuations in prices that have occurred during the last 10–15 years and have stressed the great uncertainty that exists in energy markets. First, this has probably created more humbleness with respect to the task of predicting future developments. Secondly, there is increasing focus both among theorists and planners upon subject decision making under uncertainty, i.e. procedures and strategies for how to deal with large fluctuations in market conditions.

The objective of this volume is to present a number of modelling efforts and analyses of some of the recent developments in energy markets and to discuss implications and problems for decision making and planning. Although the authors are Norwegian and primarily study energy markets and energy decisions from a Norwegian point of view, most of the contributions will be relevant for other petroleum-producing countries as well. The analytical approaches applied in several of the articles should have a wider interest for economists and people working with energy problems.

The book is divided into three parts. In Part One the focus is on the market for natural gas in Western Europe, the market structure, future development and organization. For Norway this market is of particular importance, since most of the remaining petroleum resources consist of natural gas, and Norway is one of the few major suppliers to the European market. One of the big issues at present is the prospects and effects of deregulating transmission and distribution of gas in Europe. This topic is discussed by Bjerkholt, Gjelsvik and Olsen in Chapter 1. The analysis indicates a big potential for future growth in gas demand provided that prices are brought down to more competitive levels. Deregulation will provide more direct links between producers and users of natural gas and less controlling power for the intermediary transmission companies that today dominate the European gas market. The competitive battle between the three great suppliers on the fringes of the Continent, Norway, the USSR and Algeria, will come more into the open. The article provides a game theoretically based resolution of the outcome of this battle. As is pointed out in Chapter 2 – by Bartlett, Olsen and Strøm – the growth potential of natural gas may be particularly great in the residential sector, where during recent years natural gas has frequently been chosen as the primary heating source in new dwellings. The analysis in this article is based on recently developed advanced econometric techniques for representing the decision structure of the energy user.

A characteristic feature of the European natural gas market is that it is dominated by a few agents both on the demand side and the supply side. Gas is typically traded on a bilateral basis, and governed by long-term contracts between seller and buyer (a transmission company). Thus, when analysing market structure elements of oligopolistic and strategic behaviour should be taken explicitly into account. Such game-theoretical aspects are discussed in Chapter 3 by Hoel, Holtsmark and Vislie who analyse the bargaining situation in a market where there is more than one seller and/or buyer of natural gas. One of the most striking features of the empirical model is the wide range of possible prices for the players that is consistent with the well-known 'core' of the game.

What will happen when deregulating a market where negotiations play a crucial role? This question is discussed by Vislie in Chapter 4, where account is also taken of a recommendation proposed by the IEA, saying that the USSR's market share of European gas consumption should not exceed 30%. Negotiations between sellers and buyers of gas at the various stages yield a price structure which will favour the high-cost seller (Norway). Deregulation is analysed in two steps. First, downstream companies are vertically integrated in order to meet the objectives of the end users. Secondly, the market share requirement is eliminated by opening up for increased competition among upstream suppliers. The first step towards deregulation will favour the high-cost seller and the consumers, whereas competition in the upstream industry will be to the disadvantage of Norway, with an ambiguous impact on consumers' welfare, whereas the USSR will gain.

Chapter 5 by Alfsen, Lorentsen and Nyborg takes up one of the most pressing topics in today's energy debate, namely the environmental consequences of burning fossil fuels. More precisely, the chapter raises the question of what will be the environmental impact if there is a transition from the use of coal and fuel oil to natural gas in Europe. Different scenarios regarding the composition of energy demand are analysed and the emissions to the atmosphere of sulphur dioxide and carbon dioxide are derived in the various cases. Altogether, the analysis confirms that environmental concerns, when followed by proper incentives for producers and consumers, may give a big push for more extensive use of gas in Europe.

Part Two of the book includes a selection of contributions discussing optimal management of energy resources and the interactions between the energy sector and the rest of economic development. Aslaksen, Brekke, Johnsen and Aaheim (Chapter 6) discuss principles for measurement of petroleum wealth and its relations to capital figures in the national accounts. The main focus is on the aspect of uncertainty. With respect to the question of whether spending of incomes has been too high, the chapter stresses that for an evaluation of economic policy one should look at the price expectations held by the government at the time decisions were taken.

A more detailed analysis of the economic policy pursued in Norway in the period 1976–86 is carried out in Chapter 7 by Cappelen and Gjelsvik. By inter-country comparisons, the authors review and evaluate the Norwegian counter-cyclical policy in the 1970s that was based on expected future petroleum incomes. By carrying out historical simulations on a macroeconomic model, alternative scenarios for the Norwegian economy and the use of petroleum incomes are presented. Actual spending of oil incomes is found to be an intermediate case between a scenario where the incomes are transferred to an 'oil fund' and a scenario where one consumes the return on the estimated 'petroleum wealth' (as estimated in Chapter 6).

Chapter 8 by Golombek and Hoel is an analysis of optimal extraction of Norwegian natural gas when gas can be exported and used in Norway, as input in energy-intensive industries or in gas-fired thermal plants. The main purpose of the analysis is to derive a rough estimate of the resource rent for Norwegian natural gas within a traditional Hotelling type of model. The calculations yield a resource rent for 1990 significantly different from zero.

Discount rates play a very important role in decisions on when and how to deplete exhaustible energy reservoirs. In Chapter 9 by Lund, the question is discussed of which discount rates are the appropriate ones to use by a government deciding on petroleum resources amounting to a considerable part of the national wealth. Lund takes issue with the common argument (originally stated by Arrow and Lind) that the government should behave in a risk-neutral way as it is big and has many projects, and concludes with a clear recommendation of how a government should proceed.

Chapter 10 by Kobila presents a formal analysis of the choice between irreversible hydro investments and reversible thermal (gas) generation taking into consideration both uncertainty in the gas price and fluctuations in electricity demand. In the case of price uncertainty, due to the irreversibility of hydro power projects, the lesson to the planners is not to rush into hydro investments even though this may seem advantageous by simply comparing expected costs.

Some petroleum reserves consist of oil and associated natural gas. An interesting question is to consider the optimal oil extraction plan for a country that is a price taker in the oil market, when associated gas is sold according to a long-term contract determined in negotiations between the gas seller and a large buyer. This issue is analysed by Vislie in Chapter 11. It is shown that the extraction plan for oil might be significantly altered, due to the contractual requirement for gas deliveries, as compared with the optimal extraction path when no gas contract is established. Furthermore, we find that the two activities interact both in the delivery path and the equilibrium price path for natural gas.

Part Three of the book contains two chapters that are quite different in character, although the development of the oil price plays a crucial role in both. Chapter 12 by Berger, Bjerkholt and Olsen surveys possible scenarios for the international oil market with a focus on the role of OPEC and the recent trend of cooperation between OPEC and other independent oil-producing countries. From non-OPEC oil producing countries' point of view, 'a worst case scenario' is clearly one where OPEC breaks down, in which the oil price is likely to plunge. To prevent such an outcome, non-OPEC oil producers may seek some sort of cooperation or tacit agreement with the organization. The presented model simulations suggest that for Norway the income loss ('insurance premium') from held-back production in this scenario is moderate compared to the huge income losses that will result in an OPEC breakdown scenario.

Finally, Chapter 13 by Mork, Mysen and Olsen is an empirical study of the correlations between oil price movements and GDP fluctuations for a selection of countries. Both in 1973–74 and 1979–80 the jumps in oil prices had significant negative effects on the world economy. An interesting question is whether one can detect any similar positive impact on activity levels following the sharp fall in oil prices in 1986. However, empirical studies seem to indicate asymmetric responses to oil price changes. In the present study, the most evident correlation between the oil price and GDP is found for the USA, which also shows signs of asymmetric responses.

A book like this is a result of cooperation and collective efforts. The editors wish to thank colleagues – too many to mention – for help and assistance in this work. Thanks go particularly to Kari Anne Lysell, Anne Strandli and Inger Johanne Widding who have made vital contributions to the text processing and drawing of graphs.

Part One

The European Gas Market

1

The Western European gas market: deregulation and supply competition

OLAV BJERKHOLT, EYSTEIN GJELSVIK

AND

ØYSTEIN OLSEN

1.1 INTRODUCTION

'Natural gas is likely to remain an underexploited fuel from the strict perspective of economic efficiency.'
M. A. Adelman et al. (1986)

Over a period of 20 years natural gas has become one of the major sources of energy supply for European households, business and utilities. The overall share of natural gas in the energy use in Europe has increased from somewhat above 3% in 1966 to just over 15% in 1986. Whether this expansion should be considered as fast or slow is a contested issue. According to critical observers such as, for example, Odell (1988) and Adelman *et al.* (1986), the expansion has been far too slow as a combined result of unrealistic pricing policies of the producing companies, monopolistic practices in transmission and distribution, misperception of the natural gas supply situation in Europe and various institutional constraints.

For the future role of natural gas in the energy supply of Western Europe the immediate years to come may be of particularly great importance for the role of gas far into the next century. The big issue is deregulation, but it is not the only matter of importance.

On the supply side there is a bargaining battle coming up between the three contenders Algeria, Norway and the USSR about the replacement of Dutch exports and decreasing indigenous gas reserves. On the demand side, there may be more countries hooked on to the main transmission grid in Europe and there will be more customers connected to the distribution network in the major gas-consuming countries.

The increased awareness of environmental risks may become a factor which will work strongly in favour of natural gas which is cleaner than its closest substitutes fuel, oil and coal. Natural gas may also replace nuclear power for environmental reasons, accentuated by the Chernobyl disaster. Technological development in cogeneration and other energy-using equipment may also work to promote natural gas as the preferred choice on economic as well as environmental grounds.

The deregulation issue has emerged with two major references: the deregulation of natural gas markets in North America and the intention of the Commission of the European Community (EC) to remove all obstacles to free trade within the EC by 1992. The possibility of deregulation has caused some consternation in the transmission and distribution companies. At the present time, there is more bewilderment than anything else about what the consequences of the EC's intention will be for the natural gas markets. For the supplying nations outside the EC, deregulation of gas markets may have great economic importance and influence the producing companies' ability to capture the various kinds of rents inherent in gas markets.

The main purpose of this study is to investigate possible effects of a deregulation of the European gas market, that is, introduction of the principle of common carriage or open access to the European transmission and distribution system. After a brief description of the European gas market (section 1.2), we analyse the role of the transmission companies in the light of static welfare economics (section 1.3). After a brief discussion of the common carrier principle, we present in section 1.5 a price/netback analysis of various market segments. These calculations reveal that, at least in some countries, transmission companies have exploited monopoly power, and thus restricted gas consumption. The effects on demand of a non-profit pricing policy in transmission is then compared with the prevailing pricing policy. The simulations indicate that in a deregulated market, gas consumption may increase by 13–20% in major consuming countries.

The final section takes this analysis further and studies possible supply responses initiated by common carriage. This is done by simulating a dynamic oligopoly model for the European gas market. These model runs predict a growth in gas consumption between 47 and 80% from the base year to 2010. This demonstrates that consumers will benefit from the introduction of common carriage.

1.2 THE EUROPEAN GAS MARKET

'... the mere five-point increase in gas' percentage contribution to the energy market over the past decade and a half represents a failure by the gas industry and government energy policy makers to accept the opportunities offered by natural gas for changing Western Europe's energy system.' P. Odell (1988)

1.2.1 The evolution of a European natural gas market

The discoveries of significant indigenous gas reserves, first in the Netherlands and other continental countries and later on in the North Sea, along with large suppliers made available by the USSR and Algeria, have allowed a gradual evolution in gas consumption in Western Europe. Due to high, but declining average costs of transportation, natural gas penetrated first to electricity-generating utilities and large energy-intensive industrial plants (Table 1.1). However, as the gas distribution network was expanded, natural gas accelerated as a primary fuel chosen by households and in smaller industries and the commercial sector as well.

During the 1970s, gas demand increased rapidly in the major countries specified in the table. In the aftermath of the two oil price increases, energy consumption in the European countries has stagnated or decreased in the 1980s. Natural gas, however, has continued to penetrate the energy market, although at a lower pace than in the preceding decade. The share of gas in total energy demand thus rose from about 14.5% in 1980 to 15.2% in 1986. The growth in gas consumption remained strong in Italy and the UK whereas demand levelled out in West Germany and France. Some smaller countries added to total gas demand.

In recent years, natural gas has had greatest success in the residential and commercial sectors (Table 1.1). In 1986, gas consumption in these sectors constituted about 53% of total gas demand in Europe, while this share was only 37% in 1975. Ease of control and high efficiency, in particular in central heating systems, have motivated households to switch to natural gas, both through conversion and retrofit investments. Even more important, however, has been the tendency of installing gas in new dwellings. In several countries the share of gas-heated dwellings among new homes has come to exceed 50%, and was in the range of 70–80% in the UK in the mid-1980s.

In the industrial sector and in power generating, gas consumption stagnated in the 1980s (Table 1.1). This has been due to a generally low activity level in this period, energy conservation and changes in the industrial structure. In addition, pricing policies in several countries have been directed to encourage extensive use of domestically produced coal and nuclear energy.

Table 1.1 Natural gas consumption, Western Europe, million tonnes oil equivalent (mtoe)

	1965	1970	1975	1980	1986	Average growth 1980–86
Austria	1.6	2.5	3.6	4.2	4.4	0.8
Belgium and Luxembourg	0.1	3.5	8.3	9.1	6.7	−5.0
Denmark	0.0	0.0	0.0	0.0	1.0	
Finland	0.0	0.0	0.7	0.8	1.0	3.8
France	5.0	8.4	17.1	21.9	24.8	2.1
Greece	0.0	0.0	0.0	0.0	0.1	
Iceland	0.0	0.0	0.0	0.0	0.0	
Republic of Ireland	0.0	0.0	0.0	0.5	1.1	14.0
Italy	7.3	10.8	18.6	23.1	28.9	3.8
Netherlands	1.6	15.7	32.0	31.0	33.1	1.1
Norway	0.0	0.0	0.0	0.0	0.0	
Portugal	0.0	0.0	0.0	0.0	0.0	
Spain	0.0	0.1	1.3	1.8	2.5	5.6
Sweden	0.0	0.0	0.0	0.0	0.2	
Switzerland	0.0	0.0	0.5	0.8	0.9	2.0
Turkey	0.0	0.0	0.0	0.0	0.3	
UK	0.8	10.4	32.1	41.1	48.3	2.7
West Germany	2.5	12.8	35.0	43.3	41.8	−0.6
Total Western Europe	18.9	64.1	149.2	177.6	195.1	1.6
of which:						
Electricity generation		11.5	32.5	25.1	24.1	−0.7
Industry		31.4	59.4	68.2	63.5	−1.2
Residential, commercial		28.2	55.1	79.6	99.2	3.7
As share of primary energy consumption (%)		2.3	6.7	13.2	14.5	15.2

Sources: BP Statistical Review of World Energy, OECD Energy Balances.

On the supply side, some of the gas-consuming countries have significant domestic gas resources of their own, but with the Netherlands as the only net exporter (Table 1.2). Three main producing areas supply the region from its fringes, namely the USSR, Norway and Algeria. The USSR has close to 40% of the total reserves of natural gas in the world. Algeria's exports consists partly of piped gas to Italy, and partly of LNG (liquified natural gas) deliveries to several countries on the Continent. Norway's offshore production of natural gas increased rapidly in the 1970s and all its production (close to 30 bcm in 1987) is exported to the UK and the European continent through pipelines.

Table 1.2 Natural gas reserves and production, 1986

	Production (bcm)	Proved reserves (1000 bcm)	R/P ratio	Net exports to W. Europe
France	3.60	0.04	11.11	−19.93
Italy	12.93	0.30	23.20	−16.42
Netherlands	57.03	1.80	31.56	23.90
Norway	27.30	3.00	109.89	27.30
UK	38.27	0.60	15.68	−10.04
West Germany	11.29	0.20	17.71	−30.53
Others	4.60	0.26	56.52	−21.80
Total Western Europe	155.02	6.20	39.99	−47.52
Algeria	42.10	3.00	71.26	24.60
USSR	733.80	41.10	56.01	38.80

bcm = billion cubic metres, 10^9 m^3.
Sources: BP Review of World Gas, OECD Energy Balances.

Altogether, the supply situation for the European natural gas market seems abundant. The consuming countries are connected to four large supply regions: Groningen in the Netherlands, the Algerian Sahara, Uringoi in western Siberia and the North Sea. The gas reserves included in these fields represent potential for many years with total consumption at a considerably higher level than today's. Moreover, most of the major countries in Europe are interconnected in a central transmission system. Still, it is the conventional view among many analysts that the future growth in gas consumption in Europe will be moderate (the projections range from a decline to a modest increase in total consumption, see e.g. Bjerkholt, Gjelsvik and Olsen, 1989). The background for this somewhat pessimistic picture may be found in the existing structure of the gas market, and in particular in the strategies pursued by the transmission and distribution companies.

1.2.2 A brief overview of the prevailing market structure

Natural gas is an **exhaustible resource** which means that the cost of production includes, in addition to the factor cost of bringing it to the wellhead, an opportunity cost of reducing the amount that can be produced in the future. This opportunity cost is the rationale of a resource rent to be included in the marginal cost. (For a further discussion of this aspect, see Golombek and Hoel, 1989.) Another important feature is that increased production over existing capacities will typically be made available by large-scale investments in development of new fields. There is thus lumpiness on the supply side. As will be discussed in section 1.6, these technological features may have significant effects on market behaviour.

Gas at the wellhead is still far from the end user. The transportation of natural gas in Europe is undertaken by pipelines, first from wellheads to import terminals, then through national transmission grids and, finally, via local distribution networks to the final end users.

The cost components of natural gas thus consist of extraction (production), transportation from wellhead to import terminals, national transmission and local distribution.[1] In Table 1.3, cost estimates for a number of natural gas fields serving Europe are reported. The cost estimates vary over a large range, probably due to different assumptions on uncertain parameters such as investment costs, depletion rates/production capacities, reserve estimates, etc. Distribution and transmission costs are the dominating cost components.[2]

For the inexpensive and close-to-market Groningen field, distribution costs exceed 90% of the total. For the 'high-cost' Troll field, extraction constitutes around 20%, international transport around 15%, and distribution the residual 65% of total costs. Even for LNG exports, where costs of liquefying, shipment and regasification are more than double of average international pipe transportation costs, and for gas shipped from the permafrost area of Urengoi in Siberia, distribution costs exceed 50%.

The transmission lines and local distribution networks have the same lumpiness and indivisibility properties as the production capacity. A transmission and distribution network to serve a given set of end users will for this reason often have spare capacity. Increased demand may thus imply lower, rather than higher, average transportation costs per unit. Investments in new transport capacity to cater for even higher demand may also imply lower average unit costs as better use may be made of the already existing infrastructure.

Economies of scale in transportation of natural gas imply that there will typically be a limited number of distribution companies serving each

---

[1] The distribution cost figures presented in Table 1.3 include local distribution, storage facilities to handle peak load demand and national transmission costs.
[2] It is assumed that small-scale consumers pay transmission and distribution costs while large-scale consumers only pay transmission costs.

Table 1.3 Production, transport and distribution costs (1984 $US per million BTU)

	Norway				Soviet				Algeria				Netherlands	
	Ekofisk		Sleipner		Troll		Urengoi		Pipe		LNG		Groningen	
	Low	High	Low	High	Low	High	Low	High	Low	High	Low	High	Low	High
Production	1.00	1.00	1.08	3.14	0.81	1.73	0.42	0.61	0.05	0.28	0.05	0.28	0.01	0.31
Transport[a]	0.50	0.64	0.55	0.85	0.66	1.36	1.03	2.65	1.46	1.84	1.96	2.73	0.12	0.18
Distribution:														
domestic	2.92	4.00	2.92	4.00	2.92	4.00	2.92	4.00	2.92	4.00	2.92	4.00	2.92	4.00
large scale	0.58	0.87	0.58	0.87	0.58	0.87	0.58	0.87	0.58	0.87	0.58	0.87	0.58	0.87
Total unit cost:														
domestic	4.42	5.64	4.55	7.99	4.39	7.09	4.37	7.26	4.43	6.13	4.93	7.01	3.05	4.49
large scale	2.08	2.51	2.22	4.86	2.05	3.96	2.03	4.13	2.09	3.00	2.59	3.88	0.71	1.36
C.I.F.[b] unit cost	1.50	1.64	1.63	3.99	1.47	3.09	1.45	3.26	1.51	2.13	2.01	3.01	0.13	0.49

[a] Transport costs to a central point of the European gas market.
[b] Cost, insurance, freight.
Sources: Adelman and Lynch (1986); Dahl and Gjelsvik (1988); Messner and Strubegger (1986).

market. Moreover, the transmission of gas from wellheads or import terminals to local distribution companies is by and large undertaken by transmission companies of which there are only few altogether and each of which is a virtual monopoly in its region. The major transmission companies thus have a key role in the gas market.[3]

The transmission companies may be regarded as natural monopolies. They are in a strong position *vis-à-vis* the producing companies and control almost completely the access to end users. The transmission companies are also tied together to some extent through joint ownership. On the other hand most of them have been organized on a public utility basis with government participation or working within the limits of government concessions. Thus, they may not exploit their monopoly position to the limit.

1.3 STATIC EQUILIBRIUM THEORY APPLIED TO THE EUROPEAN GAS MARKET

In a competitive market of an 'ordinary' good the equilibrium price is defined by the intersection of a downward-sloping demand curve representing the aggregate marginal utility schedule of many small consumers and an upward-sloping supply curve representing the aggregate marginal cost of production schedule of many small producers. The equilibrium price is thus equal to marginal cost. The total net benefit accruing from the consumption of the good is split between 'consumers' surplus' falling to the consumers and the 'profit' of intramarginal producers. In a long-term equilibrium intramarginal profit is eliminated by competition and the market equilibrum is depicted by the well-known textbook Fig. 1.1 showing the intersection of the demand curve (D), the marginal cost curve (MC) and the average cost curve (AC). By the fundamental theorem of welfare theory the competitive equilibrium is sufficient for efficient allocation of resources in the absence of externalities (in a wide sense), but only if all other markets are in similar equilibrium. We take these textbook commonplaces as our starting-point in discussing the peculiarities of gas production and trade in Western Europe. The European gas market differs in almost every respect from the textbook paradigm except that also in the gas market there are a large number of small consumers.

Let us point out the differences of greatest importance for our discussion with reference to the overview of the gas market given in section 1.2. As

---

[3]In some countries, like the Netherlands and Belgium, there is one company controlling the national gas transmission network. In West Germany there are eight regions with eight different transmission companies. The largest, Ruhrgas, has shares in three of the others, and more important, it controls the national transmission network. This makes Ruhrgas a dominant firm in West Germany. (For a more detailed overview, see Bundgaard-Sørensen and Hopper, 1988.)

Static equilibrium theory applied to the European gas market

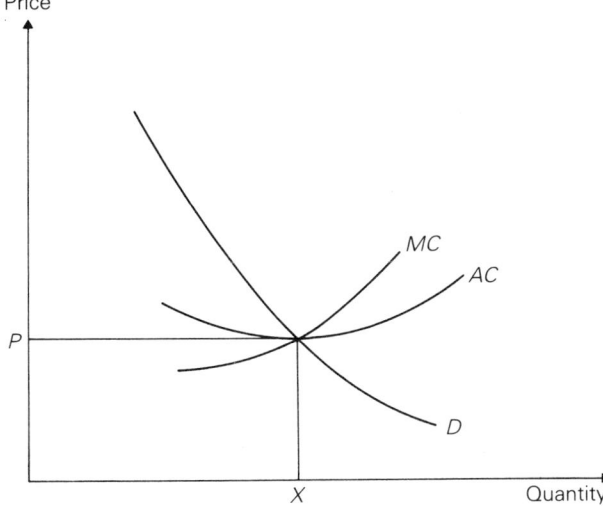

Fig. 1.1 The long-term equilibrium of a competitive market for reproducible goods.

mentioned there, due to resource scarcity in the supply of natural gas different producers will have differing marginal costs, not only as a transient phenomenon to be eliminated by competition, but as a permanent feature. Intramarginal profit will thus not be eliminated by competition. Secondly, increased production over existing production capacities will typically be made available by large-scale investments in development of new fields. And thirdly, the number of producers is relatively small, which raises the question of imperfect competition, i.e. the producers' ability to capture more than their fair share – as defined by perfect competition – of the total value of the gas produced.

The transportation and retailing of natural gas are again very different from the corresponding services of 'ordinary' goods, for which these aspects are usually left out of the analysis of market equilibrium, as being rather inessential. Technologically, we have the textbook case of a 'natural monopoly', i.e. downward-sloping average cost curves in the distribution of natural gas. Increasing returns may be caused by underutilization of capacity because of indivisibility, by lumpiness of investments as new projects are large relative to the size of the market, or by other technological reasons. The increasing returns in distribution could even outweigh decreasing returns in production. The existence of increasing returns to scale is obviously of major importance for the present state of the market. The end users are in practice constrained to purchasing from only one company, they have no possibility of storing the commodity and have thus no way of counteracting price discrimination between end users.

The specific features of the gas market lead to various kind of rents. We have already mentioned the **resource rent** accruing from the exhaustible nature of gas resources. The small number of agents producing and trading gas together with the elements of increasing returns may lead to **monopoly rent**. (If increasing returns prevail, one may have monopoly rent with zero profit.) Finally, the lack of arbitrage possibilities for end users allows **rent from price discrimination**, which in principle could amount to capturing the entire consumers' surplus.

We shall illustrate the equilibrium in the gas market in a static stylized setting, with competitive conditions in production, but with a transmission monopoly. In Fig. 1.2, the MC curve represents total marginal cost and the MPC curve marginal production cost. Marginal transmission cost (not drawn) is hence represented by the difference between the two curves. MPC is everywhere increasing. Where MC is less steep than MPC there are increasing returns in transmission. The curve AC is the sum of MPC and the average cost of transmission. The vertical distance between AC and MPC diminishes, which means increasing returns to scale in transmission of gas. The demand curve is D, while the corresponding marginal revenue is indicated by the curve MR. For simplicity it is assumed in the figure that the transmission company buys gas at competitive conditions from producers. The optimal sales volume from an overall efficiency point of view is at the intersection of the marginal cost curve and the demand curve.

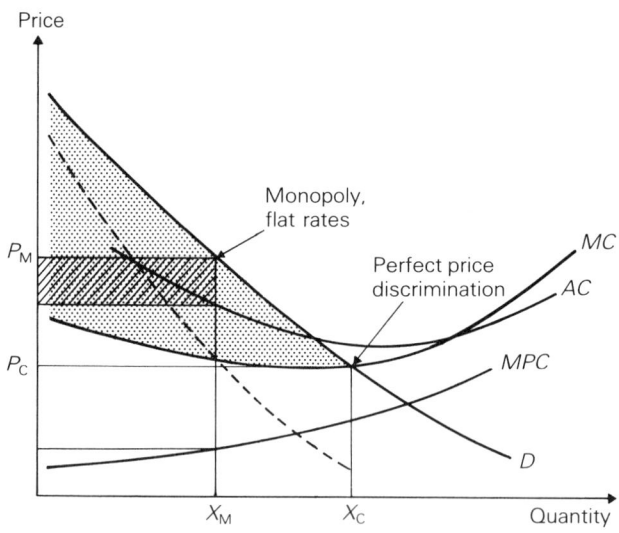

Fig. 1.2 Market solutions for a transmission monopoly.

A transmission monopoly free to discriminate between end users (discriminating monopoly) will in theory generate exactly this solution. The tariff structure can be designed to capture the entire consumers' surplus (gas should be offered to consumers at declining tariffs corresponding exactly to their willingness to pay). The tariff will charge each end user the same optimal marginal rate. This holds whether the monopoly owns the gas it transports or not. This hypothetical situation is depicted in Fig. 1.2, yielding the equilibrium quantity X_C. Since the efficiency volume is realized, there is thus a certain rationale in an unconstrained transmission monopoly. The marginal price paid by the consumers is P_C, while the average price is higher. Compared to the competitive equilibrium, the consumer surplus (the shaded area) is transferred to the transmission companies.

In practice, a transmission monopoly will not be able to apply a perfect discrimination of end users. It will rely on flat rates, at least for larger market segments, and take its profit from monopolistic rates and restricted volumes rather than perfect discrimination. The other extreme is thus no discrimination, but monopolistic tariffs. This situation, which is that of a textbook monopoly, is also depicted in Fig. 1.2. The monopoly solution has price P_M and sales volume X_M. In a fully exploited monopoly situation as drawn here, the average cost of transmission is smaller than the transmission companies' margin, which means excess capacity and a positive monopoly rent.

Most probably, the actual market equilibrium in the gas market is somewhere between the two extreme solutions. Since we do not have full information of the cost and demand structure in the market, we cannot know how close the present market solution is to the pure monopoly case, X_M. The degree of distortion and monopoly power in the market is an empirical question to which we will turn in section 1.5. First, however, we will discuss the new suggested 'order' for selling and distributing natural gas, known as 'common carriage'.

1.4 THE COMMON CARRIER PRINCIPLE

A repeated issue in recent discussions of the European gas market is the need for major structural changes in institutions and contractual arrangements. The discussion has been spurred by recent developments in the North American gas market and also by confronting the current market structure with the trade principles of the EC. The latter aspect has been emphasized, in particular, with reference to the intention of the EC Commission to remove all trade obstacles and bring the open market into full effect by 1992. Applications of these principles to the natural gas markets have been referred to as 'open access', 'common carriage' or 'common carrier principle'. But what is the 'common carrier principle' and how will it affect the gas market in Europe?[4]

Critical observers of the European market have for many years argued that the prevailing market structure and market arrangements allow for exorbitant rents both to producers and to transmission companies, and that the high end-user price that follows has severely limited the expansion of the market and resulted in underutilization of transmission capacity. A recent statement by a long-time critic is Odell (1988) who blames the 'club' of companies led by people of 'limited horizons':

> Thus in the Western Europe gas market today there is the double irony of under-exploited supply potential and an underdeveloped market. The misconceptions over gas supply and gas markets are, moreover, not simply allowed to persist by the powerful club of gas transmission and distribution companies/institutions (some state and some private). They are deliberately encouraged by them .... Their management principal objective ... appears to be to find guaranteed long term supplies just adequate to meet their predetermined calculations of markets which have been chosen in such a way that they do not have to worry much at all about competition from alternative energy sources. The strategy overall reflects a 'satisficing' approach by management which is anxious to be seen doing a technically excellent job, but which has no stomach to accept the challenges and to respond to the opportunities of a competitive approach to Western Europe.

The discussions of reform centred on the common carrier principle have received strong and articulate opposition from the 'club' members. Ruhrgas board member B. Bergmann (1988) argues that 'the current healthy state of the European gas markets is due to careful long term planning and financing by national gas monopolies and large integrated companies, and that enforced common carriage would wreak havoc with gas company planning'. Another statement from a similar source says that 'the present system and gas supply in Europe is sufficient and that any move to modify the present structure by introducing throughout Europe a blanket obligation on gas companies to transmit gas for third parties would undermine security of supply, cause uncertainties in the market, and be detrimental to the interests of the end users'.

The vehement reaction of the transmission companies towards a change in the rules of the game in the direction of common carriage is embedded in a set of arguments of why the North American development cannot be applied to Europe. It is argued that the common carrier principle is

---

[4]There may be a distinction between 'common carriage' and 'open access', the first implying an absolute obligation to carry a shipment of gas, while the latter is the weaker obligation to carry a shipment in the case of idle capacity on a first-come first-serve basis. In this chapter we use the two expressions synonymously for the weaker obligation.

incompatible with the current reliance on take-or-pay import contracts and that a change in existing contracts cannot be enforced because in Europe there is no authority corresponding to the Federal Energy Regulatory Commission able to exert regulatory powers over all parties. Furthermore, it is argued both that common carriage would endanger the energy supply security of Western Europe and that common carriage would leave small distribution companies as easy victims of take-overs and thus result in strengthened monopoly/monopsony power rather than the opposite.

The position of the EC Commission as to what common carriage means and how it should be implemented for the gas market is by no means clear. Various documents by the European Commission describe and discuss the problems concerned, which may be outlined as follows:

1. Harmonization of taxes and prices and the obligation to publish distribution tariffs and prices of individual contracts (price transparency).
2. Open access to the national and international pipeline systems, i.e. the obligation to allow the gas suppliers to carry any volume of gas to a 'third party' (end user). This principle of common carriage means the end of a system of national monopoly for gas transmission companies as retailers of gas.
3. Abolition of prevailing obstacles to free competition between different fuels for electricity generation. Important is the banning of gas burning and protection of nuclear and domestic coal in some markets.

The EC has not come out with an official position in these matters yet, but several documents indicate that EC officials are leaning in favour of the common carrier principle and other measures to promote competition in European energy markets. While efficiency considerations and the general principles of the EC clearly favour reform of the prevailing market structure, it is still difficult to guess the final outcome of the political handling of this problem within the EC. The political authorities of the EC have not shown a strong interest in promoting competitive energy markets until recently. The EC Council has *inter alia* prohibited any further use of natural gas in government-owned power plants. Underlying this regulation, which clearly has to go if gas markets are to become more competitive, is the protection of domestic energy sources in the respective EC countries: coal in the UK and West Germany and nuclear power in France and Belgium.

Consumers, independent producers and regulators have common interests in looking for policy means to enforce more efficient ways of trading and transporting natural gas within Europe. For Statoil and other large producers it seems clear that they do not have the same interest in holding on to the take-or-pay contracts as the transmission companies. The lower prices since 1986 have created greater interest in raising additional revenue by finding outlets for supplementary gas resources, maybe to the extent of trading the apparent security of the take-or-pay contracts against the possibilities of direct contracting with end users and local distribution companies opened up by the common carrier principle. High-cost producers with fields that would not have been developed under more competitive conditions, stand to lose in a more competitive market, however. Statoil's position is clearly vulnerable from a cost point of view, especially if *Glasnost* thaws away any political limit set on Soviet supplies. Statoil's control over the entire Norwegian production and transportation to the Continent and the UK gives the company considerable flexibility in its marketing.

1.4.1 A theoretical definition of common carriage

Let us return to the simplified theoretical scheme from the previous section. The common carrier principle can be taken to mean access to the use of the transmission pipeline at current average costs, i.e. the costs corresponding to the volume X_M in Fig. 1.2. Producers and end users would then have a margin of mutually beneficial trades. Market forces could then be relied on to bringing the end-user price down until it equals AC. This new situation is depicted in Fig. 1.3, where the 'common carrier' equilibrium is given by the price P_{CC} and sales volume X_{CC}. This is still a higher price and lower volume than the (unobtainable) competitive equilibrium given by price P_C and volume X_C. The main point is, however, that the move from P_M to P_{CC} reduces transmission costs. The transmission companies' surplus vanishes, and the transport tariff is reduced to average transmission costs.

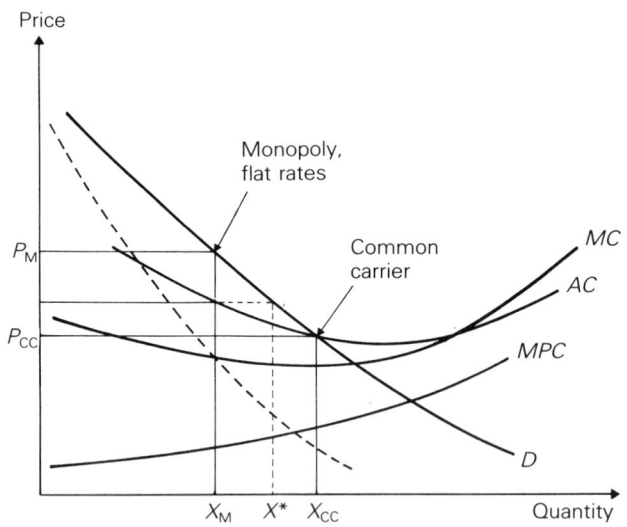

Fig. 1.3 Market solution with common carrier.

Based on this stylized theoretical framework we conclude that common carriage may lead to higher volumes traded and more competitive behaviour in the gas market (see also the discussion in Hurst, 1988). There are, however, other ways of regulating the transmission monopoly. One is by forcing the transmission monopoly to set its rates on a traditional public utility cost-of-service basis. This will result in pipeline tariffs set at average costs and theoretically lead to the same result as common carriage. It may, however, imply a stronger bias of a too capital intensive transmission system and also incur higher administrative costs. Another way is by direct regulation of end-user prices towards the same equilibrium solution. The latter alternative combined with common carriage may be the right remedy to speed up adjustment in a transient phase of an underdeveloped market. Increasing oil prices as might result from the recent OPEC accord might help in this respect.

What then about the claim that common carriage will simply transfer monopoly power and benefits from the transmission companies to producers of natural gas? Clearly, one cannot disregard the possibility that producing companies will take advantage of the new situation and try to capture a part of the consumers' surplus by price discrimination. However, as pointed out in section 1.2, there are several potential suppliers to the European market, each having significant reserves, and in an open market there are thus reasons to expect strong competition over market shares. We will return to this in section 1.6.

Existing take-or-pay contracts may, of course, prove less profitable for producers as well, as a consequence of 'third parties' entering the market with new deliveries. In particular, contracted gas from high-cost fields may suffer from the fact that they have been developed too early.

1.5 A PRICE/NETBACK ANALYSIS OF MARKET SEGMENTS, AND THE EFFECTS OF NON-DISCRIMINATING PRICING

In this section we attempt to throw some empirical light on the profit-making behaviour of the transmission companies by presenting figures on prices paid by different consumers and cost and profit margins in the transportation of natural gas. To the extent the analysis reveals a positive netback to distribution companies and/or price discrimination, it is an indication of the use of monopoly power.

The netback analysis is carried out for West Germany, France and the UK. For each of these countries we distinguish between three market segments: households/commercial, industry on firm supply contracts and industry on interruptible contracts. Eurostat (1988) provides data on gas price for end users of different yearly off-takes in the first two sectors. For households, we have chosen a weighted price corresponding to a

125.6 GJ year^{-1} consumption level. For firm industries, the price for an off-take at 418 600 GJ year^{-1} is used. For interruptible industrial deliveries, the price of heavy fuel oil is used in the calculations.

The cost figures are collected from Purvin and Gertz (1987). There, the following cost components are specified and estimated:

1. town distribution (assumed zero for industry);
2. national transmission, including storage costs;
3. import price (c.i.f.);
4. value added tax on the components above (assumed zero for industry).

Subtracting the components (1)–(4) from the end-user price, gives the corresponding net profit margin or **netback** to the transmission company.

Estimates of distribution costs and netbacks for two years, 1984 and 1987, are reported in Table 1.4. In the same table, we present calculated 'unit cost prices' for each user group, defined as equal to the estimated total costs. The table shows quite significant netback margins and price differentials. In West Germany, the highest netbacks are calculated for firm industrial deliveries, while margins for interruptible contracts are rather small. The margins generally decreased from 1984 to 1987. End-user gas prices declined less over these years than fuel oil (premium factors increased). Thus, in spite of the squeeze in margins, gas lost competitiveness against fuel oil, especially heavy fuel oil, but closed the gap with coal prices. In the UK, netback margins in households and smaller industries were negative in 1984, but turned positive in 1987 after the import price plummeted. The pricing policy tends slightly to disfavour large industrial users, but also industry netbacks were much lower than in West Germany. The figures calculated for the UK indicate that British Gas kept prices low and let a part of the gas rent be passed on to the consumers before 1987. The decrease in gas prices in 1986–87 has been relatively less. As opposed to the other two countries, in France the estimated netbacks are highest for the group consisting of households, commercial and other smaller industries. Based on these figures, Gaz de France seems to have had a pricing policy that has favoured larger industries. The margins are at roughly the same levels both in 1984 and in 1987.

To sum up, price differentials between firm industry supplies and other uses have been large in France and the UK, but insignificant in West Germany. The German pricing policy seems to have been that of a non-discriminating monopoly squeezing all consumers at the same level, the French policy has discriminated against households and other smaller users, while the British policy has favoured and partly subsidized smaller consumers, probably at the expense of indigenous gas producers.

Obviously, various uncertainties are inherent in the estimated distribution costs (see Table 1.3). Since import prices are assumed to represent the cost of gas to the transmission companies – if gas can be bought cheaper from

Table 1.4 Netback to gas distribution by sector (1987 $US per million BTU)

	Household/commercial			Industry, firm contracts			Interruptible contracts	
	1984	1987	Av. 1980–87	1984	1987	Av. 1980–87	1984	1987
West Germany								
End-user price	12.36	7.89	11.87	9.10	5.87	8.43	7.38	3.01
VAT	1.52	0.97	1.46	0.00	0.00	0.00	0.00	0.00
Town distribution	2.34	2.34	2.34	0.00	0.00	0.00	0.00	0.00
National transmission	0.84	0.84	0.84	0.84	0.84	0.84	0.84	0.84
Import price (c.i.f.)	5.12	2.44	4.47	5.12	2.44	4.47	5.12	2.44
Netback	2.54	1.31	2.76	3.14	2.60	3.12	1.42	−0.27
Total unit cost[a]	9.46	6.40	8.72	5.96	3.28	5.31	5.96	3.28
Premium factor[b]	1.05	1.41	1.13	1.23	1.95	1.45	1.00	1.00
France								
End-user price	12.74	9.63	11.80	6.53	3.86	6.11	7.75	2.86
VAT	2.00	1.51	1.85	0.00	0.00	0.00	0.00	0.00
Town distribution	2.33	2.33	2.33	0.00	0.00	0.00	0.00	0.00
National transmission	0.87	0.87	0.87	0.87	0.87	0.87	0.87	0.87
Import price (c.i.f.)	5.33	2.18	4.23	5.33	2.18	4.23	5.33	2.18
Netback	2.21	2.74	2.52	0.33	0.81	1.01	1.55	−0.19
Total unit cost	10.12	6.38	8.81	6.20	3.05	5.10	6.20	3.05
Premium factor	0.90	1.09	0.92	0.84	1.35	1.07	1.00	1.00
UK								
End-user price	7.20	6.67	6.82	5.46	5.12	5.41	6.28	3.05
VAT	0.00	0.00	0.00	0.00	0.00	0.00	0.00	0.00
Town distribution	2.43	2.43	2.43	0.00	0.00	0.00	0.00	0.00
National transmission	0.90	0.90	0.90	0.90	0.90	0.90	0.90	0.90
Import price (c.i.f.)	4.04	2.92	3.59	4.04	2.92	3.59	4.04	2.92
Netback	−0.17	0.42	−0.10	0.51	1.30	0.93	1.33	−0.77
Total unit cost	7.37	6.25	6.92	4.94	3.82	4.49	4.94	3.82
Premium factor	0.72	1.19	0.86	0.87	1.68	1.26	1.00	1.00

[a] Total unit cost = import price + town distribution + national distribution + VAT
[b] Ratio of gas price to alternative price, here LFO in household and HFO in industry sector.
Sources: Town distribution and national transmission, Purvin and Gertz (1987); end-user price, Eurostat (1988).

indigenous producers, the margins will be underestimated. This may, for example, affect the estimates for the UK, where British Gas in periods has set indigenous prices below import prices on gas from Norway. Furthermore, a large fraction of the transmission and distribution costs are fixed capital costs. As a result, the unit costs presented in Table 1.4 vary with several economic and geographical factors in the various countries and market segments.

How can the calculated profit margins and the actual volumes of natural gas brought to the market-place be evaluated in light of the theoretical discussion in sections 1.3 and 1.4? Recall again Fig. 1.3. The average cost in transmission in the current situation generally decreases with the volume of gas transported through the pipeline. If we ignore the possibility of having a perfectly discriminating monopoly, the observed equilibrium point, for which distribution costs are calculated, is somewhere to the left of the quantity X_{CC}, which yields a zero netback in transmission. Thus, based on the calculated costs and netback margins, a main conclusion is that the volumes of gas in the markets are too small, i.e. smaller than what should have been attainable under more competitive market conditions. To evaluate the degree of distortion in the market simply on this kind of information is, however, not possible. To do this in a satisfactory way would have required full information of the various cost and demand functions, which we do not have.

What we instead intend to do in the following is to estimate the demand effects of gas prices being reduced from the actual levels to prices corresponding to the calculated total unit costs in Table 1.4. If the gas companies reduced prices to this level, gas demand would increase, but still be lower than X_{CC}. So, even though ATC decreases, the distribution companies would still earn a positive margin. Assuming simple monopoly behaviour, this pricing policy would increase demand from X_M to X^* in Fig. 1.3. To measure volume effects we have used a gas demand model for the European market developed in the Central Bureau of Statistics, Norway, called GEM. This model covers all the major gas-consuming countries in Western Europe, distinguishing in each between four sectors: households, commercial, industry and power generation. GEM has been simulated for two sets of prices: (1) average 1980–87 end-user prices, and (2) unit cost prices, as presented in Table 1.4. In simulation (1), we have used the same prices in the household and commercial sectors, while firm industry prices are applied in the manufacturing sector. Other variables have been kept constant at the 1984 level. Gas consumption in the power generation sector has been kept constant throughout the analysis. The model has been run over several years in order to include lag effects and compute long-term equilibrium solutions.

The results of these simulations are reported in Table 1.5. According to these calculations, total gas consumption in West Germany (exclusive power

Table 1.5 Volume effects of unit cost pricing. GEM simulations (mtoe)

	Households	Commercial	Industry	Total
France				
(1) Historical prices 1980–87	7.0	9.6	8.8	25.7
(2) Unit cost prices	8.7	10.5	9.6	29.1
Deviation (2 − 1)	1.7	0.9	0.8	3.4
Percentage	24.3	10.4	9.1	13.6
West Germany				
(1) Historical prices 1980–87	10.1	7.6	13.8	42.0
(2) Unit cost prices	14.5	9.0	16.2	50.3
Deviation (2 − 1)	4.4	1.4	2.4	8.3
Percentage	44.5	18.4	17.4	19.8

generation) would increase by 8 million tonnes oil equivalent (mtoe) if gas prices are decreased to unit costs. The gain is largest in the household sector, almost 4.5 mtoe, while demand in the industry sector increases with 2.5 mtoe. In the UK, the volume effects are rather insignificant, due to the small differences in actual prices and unit costs reported above. In France, the total gain of more competitive pricing is 3.5 mtoe, half of which is estimated to take place in the household sector.

It should be emphasized that the losses and gains of Table 1.5 are differences in long-term equilibrium levels, which will only occur if the price differences are sustained over a long period of time. As stressed above, our constant average cost of transmission and distribution in these calculations tend to overestimate unit cost prices at the new volumes and thus underestimate volume effects. On the other hand, there is, of course, uncertainty related to the estimated cost figures.

Still, the model runs support the argument that current price policies of the gas companies have significantly restricted consumption in countries like West Germany and France. The simulations indicate that demand in the three sectors, households, commercials and industry, in these countries could increase by 10–23% if more competitive pricing policies were adopted. Lifting the ban on the use of gas for power generation would add significantly to this prospect. Actually, several analysts foresee the best potential for gas in the latter market segment if gas was allowed to compete (Odell, 1988; Rogner, 1988). Thus, the current 15% market share of gas in Europe's energy market seems far too low.

Both the comparisons between prices and costs and the model runs indicate that gas companies do not maximize consumer surplus, but rather exploit a monopoly position to capture a positive rent. However, it is also clear that the analysis is based on some rather simplifying assumptions. One complication is caused by the fact that there are a limited number of agents on the supply side as well. A realistic description of the gas market should therefore take into consideration the game situation over contracts between large producers like the Netherlands, Norway, Algeria and the USSR on one side and the national gas companies of Western Europe on the other. The outcome of negotiations may be some kind of sharing of the total rent, arising by deducting real producer and distribution costs from end-user prices. Calculations indicate that the gas contracts are designed to share rents between producers and gas companies (Bjerkholt, Gjelsvik and Olsen, 1989).

1.6 SIMULATION OF FUTURE GAS SUPPLIES AND PRICES IN A DEREGULATED EC MARKET

Our investigation so far shows that end-user prices have not come down to competitive levels. Thus, there are strong indications that the gas markets of Western Europe are underutilized. But the argument that the prevailing market structure with long-term 'take-or-pay' commitments are necessary to ensure gas supplies in the long term, is still to be investigated. Will common carriage undermine the market, get rid of 'take-or-pay' contracts and scare off investors from high-cost gas projects like Troll? Or, will it be foolish of the EC to undermine the strong position of the gas companies, leaving the battlefield open for strong and greedy producer groups ready to form a cartel?

1.6.1 A dynamic oligopoly model

To answer these questions we have made simulations on a dynamic oligopoly model (DOM) developed in the Central Bureau of Statistics, Brekke, Gjelsvik and Vatne, (1987). The model describes a game between three large producers: Norway, Algeria and the USSR, playing on an excess demand function (total demand of the Western European continent minus indigenous production). This model simulates a deregulated market in which there is no intermediate barrier between suppliers and end users, and the producers compete directly for market shares.

The UK is kept out of the game, and the Dutch production is included in indigenous supply. Since the game is essentially an investment game between suppliers with a bundle of lumpy investments, and the Netherlands already have made most of their heavy investments, this seems to be a reasonable way to reduce the number of players which makes the model easier to handle.

Each player possesses a bundle of large, lumpy investments. In the beginning of each 5-year period they can make one or more investments, or none. The moves are made simultaneously, only previous investments are known. The investments are operative in the next period. The players maximize discounted cash flows over the horizon of seventeen 5-year periods starting in 1985. They have perfect information of demand, costs and projects and can predict the other players' best moves. The players choose their best strategies on the basis of this prediction (Nash equilibrium).

A model solution consists of a complete plan of how to act in all future periods. The plans (strategies) consist of a set of actions, contingent on previous outcomes. Thus, the solution also shows the alternative optimal strategies whenever another player deviates from the optimal strategy by, say, postponing an investment.

The model is solved by dynamic programming, and the solutions are perfect Nash equilibria.[5] The solutions of this investment game are dependent on the solution of the short-run game for given investments. For the sake of simplicity we have chosen the Bertrand price game (Bertrand, 1883). This implies full capacity utilization and lower prices compared to a short-run Cournot game.

The supply behaviour assumed by DOM is strikingly different from that of a static Cournot investment game. The static game is a one-shot game, i.e. the players cannot respond to each other's actions or moves. The Cournot equilibrium is Pareto-dominated by the collusive solution (monopoly) which, however, is not achievable when producers cannot cooperate. This is known as the 'prisoners' dilemma'. But if the game is repeated, and the players have so-called trigger strategies, the collusive solution may be an equilibrium. However, we will argue that as a description of behaviour in the market for natural gas, the collusive repeated game solution is not appropriate. The main reason is that the existence of irreversible investments gives little power to trigger strategies.

To grasp the basic implications of the existence of large and irreversible investment projects in the gas market, a model building explicitly on dynamic game theory is required. In such a game the players are perfectly aware that their current actions have important implications in future periods. If Norway decides on a large investment in period 0, this will not only increase total supply and decrease market prices, but also decrease profits on the competitors' future investments. In such a dynamic game, the states and the strategies at various points in time will depend on previous actions and outcomes in the market. In equilibrium, the players will balance the profits from discouraging other supplies by making an investment, against the profits from restricting supply by postponing the investment.

---

[5]In some cases the Nash solution is not unique, i.e. there are two or more Nash equilibria. Rationality in such situations is not easy to define. We assume the minimax solution will be chosen in this case.

1.6.2 Model simulations

(a) Basic assumptions

The price and income elasticities are averaged over those used in the GEM model referred to in previous sections. The aggregate demand function gives a lower demand in 1990 and 1995 than the International Energy Agency (IEA) and Purvin and Gertz projections for the same income and price assumptions, but a slightly higher demand in 2000 and 2010.[6] Indigenous production was 123.2 bcm in the base year 1985, and is assumed to decrease by 1.2% throughout the horizon.

In 1985, exports to the demand region were 13, 17.8 and 28.6 bcm for Norway, Algeria and the USSR respectively. Norway had initially no excess production capacity. Exports are assumed to increase to 15 bcm in 1990, but decrease later on as fields expire. Algeria had idle capacity of 4.7 bcm in the Transmed pipeline to Italy in 1985. We assume this capacity to be fully utilized by 1990. Similarly, the 27.4 bcm idle capacity in the Soviet export pipeline to Europe is assumed to be absorbed by 1990.

Each player has three possible investment projects. For Norway these are Sleipner (5 bcm) and Troll I and II with 24 bcm each. Algeria can install a compressor platform in Transmed, adding 5.5 bcm to total capacity. The second project is building another pipeline across the Mediterranean Sea, adding another 18 bcm to total export capacity. The third possibility is utilizing and restoring 'idle' LNG capacity, amounting to 20 bcm. The USSR can install extra compressors increasing existing pipeline capacity by 12 bcm. The other options are two new pipelines to Western Europe with 30 bcm capacity each.

The estimated cumulated investment costs are (in million $US):

Projects	1	2	3
Norway	1 900	8 400	14 400
Algeria	400	6 700	7 700
USSR	200	9 200	18 200

In the model, variable costs are the sum of variable production costs like labour, material, insurance and energy costs, and transport costs to a central point in the European market. For producers not investing in pipelines, transport costs are total average unit costs including capital costs. This applies to all three Norwegian projects, which are all field investments.

The base case discount rate is set at 10%.

(b) Model results

The simulation results are shown in Table 1.6. In 1985, the players' optimal decision is project 0, 1 and 1 for Norway, Algeria and the USSR

---

[6] The model has been run for a more accurate calibration to Purvin and Gertz's (1987) implicit demand function. This did not change the DOM solution.

Table 1.6 Results from the DOM simulations

Year	State			Cap. (bcm)			Cons (bcm)	Price norm	Market shares		
	Norway	Algeria	USSR	Norway	Algeria	USSR			Norway	Algeria	USSR
1985	0	0	0	13	17.8	28.6	183	1.00	0.07	0.1	0.16
1990	0	1	1	13	17.8	28.6	183	1.00	0.07	0.1	0.16
1995	1	2	1	15	28	68	233	0.60	0.06	0.12	0.29
2000	2	2	1	16	46	68	250	0.70	0.06	0.18	0.27
2005	2	2	1	40	46	68	273	0.77	0.15	0.17	0.25
2010	2	2	1	40	46	68	271	0.85	0.15	0.17	0.25
2015	3	2	1	40	46	68	270	0.93	0.15	0.17	0.25
2020	3	2	2	64	46	68	293	0.91	0.22	0.16	0.23
2025	3	2	2	56.6	46	98	314	0.90	0.18	0.15	0.31
2030	3	2	2	55.8	46	98	312	0.99	0.18	0.15	0.31
2035	3	2	2	53	46	98	308	1.11	0.17	0.15	0.32
2040	3	2	3	53	46	98	306	1.22	0.17	0.15	0.32
2045	3	2	3	53	46	128	335	1.17	0.16	0.14	0.38
2050	3	3	3	53	46	128	334	1.29	0.16	0.14	0.38
2055	3	3	3	53	66	128	352	1.30	0.15	0.19	0.36
2060	3	3	3	53	66	128	351	1.43	0.15	0.19	0.36
2060	3	3	3	53	66	128	350	1.57	0.15	0.19	0.37
2065	3	3	3	53	66	128	349	1.73	0.15	0.19	0.37

respectively. This means that Norway does not invest in this period, while Algeria and the USSR both start compressor projects. These capacities are added to 1985 supplies and excess capacities, increasing total imports from the three suppliers to 111 bcm in 1990. The price will plummet to 60% of the 1985 level at 1985 $2.34 per million BTU ($84.2 per 1000 m^3) as a result of this massive flow of gas pouring into the market.

Norway decides to start the Sleipner investment in 1990, the Troll I investment in 1995, and Troll II in 2010. Algeria initiates the pipeline investment in 1990, and the LNG investment as the last investment simply at the point of time (2045) when this action maximizes discounted cash flow. The USSR puts in the first new pipeline project in 2015 and the second in 2035. As a result of these investments, the price does not exceed the 1985 level until 2025. In 2050 all investments are productive, the game is over and the price increases at the speed set by the excess demand function.

Due to our assumption regarding the short-run price game, all existing capacities in the production and transmission system are absorbed immediately. However, the most striking result from the base run is that the market absorbs an additional 50 bcm volume growth from 1985 to 1990 even though the price is above current level. Although there is presently some growth in gas demand, this is far from the 10 bcm yearly predicted by the model. As a mirror image of the continued heavy investments in supply regions, gas continues to penetrate the continental market in the 1990s and reaches 273 bcm in 2000. Thereafter, there is no further expansion in consumption until gas from Troll II enters the market after 2010.

In Fig. 1.4, the development of natural gas consumption as projected by the DOM model is compared with the projection made by Purvin and Gertz (1987) and IEA (1988). The much stronger growth in the DOM simulation is rather remarkable; while, for example, the IEA foresees a growth in gas consumption of about 20% from 1986 to 2000, consumption in our model increases close to 50% in the same period.

In Bjerkholt, Gjelsvik and Olsen (1989) some alternative simulations are undertaken, based on changes in the assumptions regarding indigenous production in consumer countries, discount rates and the start period of the supply game. They are all characterized by strong growth in consumption – ranging from 47 to 80% from the base year to 2000. In some of the model runs, the dynamic game leads to strategic investments aimed at preventing other players' investments.

Even though the dynamic supply model is not designed for projecting the development of the European gas market in any detail, it demonstrates an important theoretical point: oligopolistic competition can lead to a fierce fight for market shares, even though there are few players. If the basic assumptions hold, that is, common carriage prevails, the players cannot cooperate, and they are fully informed about prices and costs and heavy

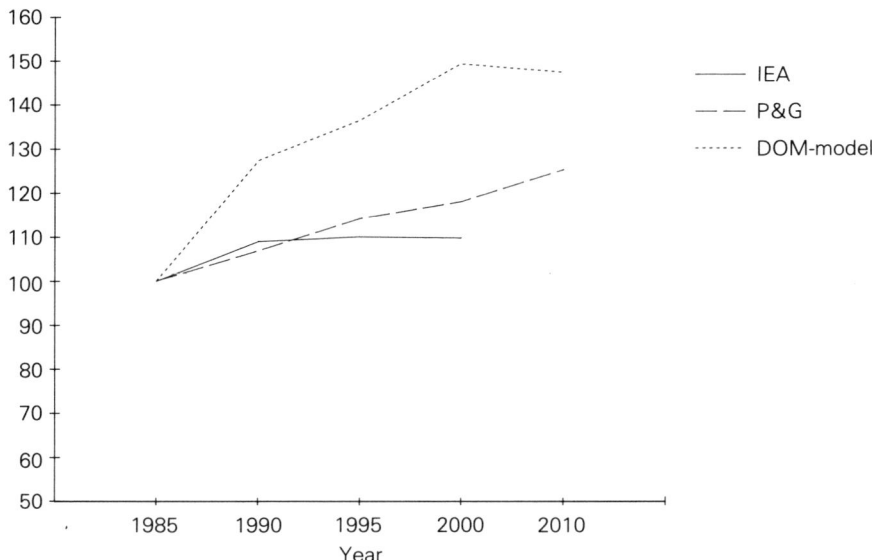

Fig. 1.4 Projections for natural gas demand, Western Europe, 1985 = 100.

investments can be financed, there seems to be little reason for worries about future supplies of natural gas to the European market if the market is deregulated. The consumers would surely benefit from it, producers will increase capacity utilization and gain market outlets for new investments, but compared to the present state of the market, they may lose rents from selling high-price gas to premium markets when oil prices are high.

The model runs are based on the assumption that the intermediate barrier between suppliers and end users is eliminated, and that the players compete directly for market shares. Clearly, a necessary condition for this assumption to hold is that a well-functioning regulatory body can be set up to ensure common carriage.

REFERENCES

Adelman, M. A., Blitzer, C. R., Cox, L. C., Lynch, M. C., Parsons, J., White, D. C. and Wright, A. (1986) *Western Europe Natural Gas Trade. Final Report*, Energy Laboratory, MIT.

Adelman, M. A. and Lynch, M. C. (1986) Natural gas trade in Western Europe: the permanent surplus, in Adelman *et al.* (1986).

Bergmann, B. (1988) The horizon for natural gas in Western Europe. *OPEC Review*, **12**(3), Autumn, 239–68.

Bertrand, J. (1883) Théorie mathématique de la richesse sociale. *Journal de Savants*, 499–508.

Bjerkholt, O., Gjelsvik, E. and Olsen, Ø. (1989) *Gas trade and demand in northwest Europe: regulation, bargaining and competition*, Discussion Paper no. 45 from the Central Bureau of Statistics, Norway.

Brekke, K. A., Gjelsvik, E. and Vatne, B. H. (1987) A dynamic supply side game applied to the European gas market, Discussion Paper no. 22 from the Central Bureau of Statistics, Norway.

Bundgaard-Sørensen, U. and Hopper, R. J. (1988) The potential for open access contract carriage in West Europe. Paper presented at the *IAEE Tenth Annual International Conference*, Luxemburg, 4–7 July.

Dahl, C. A. and Gjelsvik, E. (1989) Simulating Norwegian Troll gas prospects in a competitive spatial model, *Resources and Energy*, **11**, 35–63.

Eurostat (1988) *Gas Prices 1980–1988*, Statistical Office of the European Communities, Luxemburg.

Golombek, R. and Hoel, M. (1989) The resource rent for Norwegian natural gas, Chapter 8 in this volume.

Hurst, C. (1988) Pricing of natural gas in developing countries: some theoretical and practical considerations. *OPEC Review*, **12**(3), Autumn, 269–88.

IEA (1988) *Energy Policies and Programmes of IEA Countries. 1987 Review*, OECD/IEA, Paris.

Messner, S. and Strubegger, M. (1986) The influence of technological changes on the cost of gas supply, Working Paper -86-38, International Institute for Applied Systems Analysis, Laxenbourg, Austria.

Odell, P. (1988) The West European gas market. The current position and alternative prospects. *Energy Policy*, **16**(5), 480–93.

Purvin and Gertz (1987) *Western Europe Natural Gas Industry Market and Economic Analysis to 2010*, Purvin & Gertz, Inc., London.

Rogner, H. (1988) Technology and the prospects for natural gas. *Energy Policy*, February 9–26.

2

Residential energy demand – the evolution and future potential of natural gas in Western Europe

SARITA BARTLETT, STEINAR STRØM AND

ØYSTEIN OLSEN

2.1 INTRODUCTION

Natural gas demand in Western Europe has grown substantially since the early 1960s. In 1960, natural gas accounted for only 2% of the total primary energy consumption in the region. Initially, natural gas was used in the industrial sector, and in countries with indigenous supplies, i.e. in the Netherlands and the UK. Additional discoveries in the North Sea, and the first oil price shock in the early 1970s made natural gas more attractive as an energy supply for many potential users. Distribution first developed in countries where there were existing town gas networks, e.g. the UK and Italy. In other countries, the construction of new networks was a condition for the introduction of natural gas. Both interregional and international trade accelerated in Western Europe. In the early 1980s, natural gas demand continued to grow, but at a slower rate than in the 1970s. By 1986, natural gas represented over 15% of the primary energy consumption in Western Europe.

In the residential sector, natural gas use has grown from a very small percentage of total energy consumption in the early 1960s, to 40% of total consumption in 1985. This growth has varied among the countries in the region. In the 1970s, the increased availability of natural gas, and its favourable price relative to oil, caused fuel switching from oil to natural gas in existing dwellings, and stimulated the use of natural gas in new dwellings. The intensity of use also increased primarily because of increases in the number of centrally heated dwellings, and in the share of single-family dwellings. In the early 1980s, natural gas demand grew at a faster rate than total energy demand. Fuel switching had slowed down, but the preference for natural gas space-heating systems continued to prevail.

Will residential natural gas demand in Western Europe continue to grow in the future? One of the major uncertain factors is whether the relatively low oil prices from the mid-1980s will be sustained. Moreover, a stable economic environment is crucial for the expansion of future energy demand in general, and hence natural gas demand.

To examine the future directions of natural gas demand, a formal model is a useful tool. In this chapter we present a model which can be used to analyse residential space-heating demand. The model is estimated on data for seven European countries: France, West Germany, Italy, the Netherlands, the UK, Denmark and Sweden. We have directed our attention to the space-heating end use. This end use comprises around 80% of residential natural gas consumption. The model applied in this chapter is a dynamic, discrete–continuous choice model (Dagsvik *et al.*, 1987; Bartlett *et al.*, 1987). The discrete part of the model corresponds to the choice of fuel used for space heating in both new and existing dwellings, while the continuous part determines the level of energy demand, given the fuel choice. Transitions among fuels (fuel switching) makes the model dynamic, and a specific parameter reflects the degree of 'habit persistence' held by consumers. An important feature of this model is that the set of independent variables includes both structural and economic components. The structural variables link energy demand to the characteristics of the dwelling stock. Taking these explicitly into account the model may be regarded as a synthesis between a traditional econometric model and an engineering approach.

This chapter is divided into four sections. Section 2.2 gives a brief overview of the evolution of residential natural gas demand in Western Europe and discusses the determinants of natural gas demand, such as the number of dwellings using natural gas for space heating, and the intensity of use. In section 2.3, we present the theoretical fuel choice model. The data and estimation results are briefly surveyed. Finally, in section 2.4 we have used the model to illustrate out projections for space-heating end use.

2.2 THE EVOLUTION OF NATURAL GAS DEMAND IN WESTERN EUROPE

2.2.1 An overview

In the early 1960s, natural gas became available in the residential sector of many countries. Natural gas was viewed as an alternative to inconvenient coal-based systems. As new natural gas reserves were discovered, residential gas networks expanded, and in the late 1960s demand grew rapidly. The growth in demand was also supported by the emergence of central heating systems in Western Europe. The effects of this were twofold. First, it reinforced the households' decision to switch from coal to natural gas based heating systems, because in many countries natural gas was aggressively marketed for use in central heating systems. Secondly, the presence of central heating systems raised the intensity of use in each home. Initially, natural gas was used in countries that had domestic supplies, e.g. the Netherlands and the UK. International trade extended the market to Italy, France and West Germany. Residential natural gas consumption is shown in Table 2.1. Throughout the study period, the UK has been the largest user of natural gas, accounting for 38% of total consumption in the region in 1986.

Table 2.1 Residential gas consumption (PJ)

Country	1960	1973	1980	1986
France	52	174	291	361
West Germany	46	214	410	552
Italy	n.a.	106	233	374
Netherlands	n.a.	296	411	381
UK	144	509	888	1041
Total	n.a.	1299	2233	2709

From 1973 to 1978, natural gas consumption increased even further; the annual average growth rate was around 10%. Natural gas systems replaced oil systems in existing dwellings, and were installed in new dwellings. In most countries, the share of natural gas used to meet total space-heating demand was also growing, especially in Italy, the Netherlands and the UK. The oil price rise that occurred in 1973–74 also stimulated natural gas demand. The differential between oil and gas prices created incentives for households to replace their oil heating systems with gas systems. Figure 2.1 shows that natural gas gradually became less expensive relative to oil. Households were still switching from coal to natural gas systems.

After 1978, the natural gas demand was still increasing, but at a slower rate (except in Italy). The average growth rate for the five countries from 1979 to 1986 was 4–5%. The large increases in the price of oil in 1973–74, and again in 1979–80, caused a general stagnation in energy demand. Stagnation in national economies, central heating systems in most countries reaching saturation point, and the implementation of energy conservation policy measures designed to reduce consumption also contributed to the sluggish growth in demand. In 1979–80, the differential between the price of fuel oil and the price of natural gas was less than in

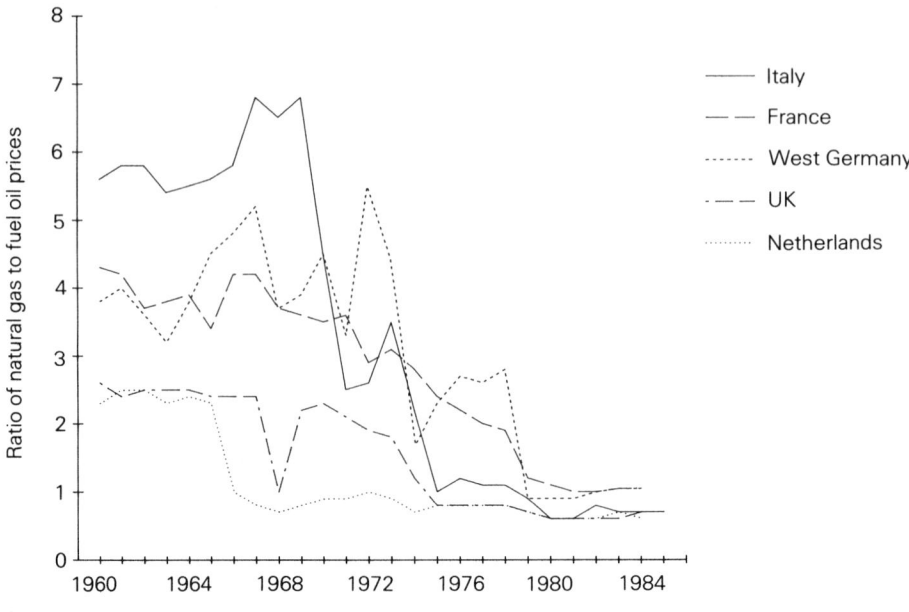

Fig. 2.1 Relative energy prices.

the previous years, and natural gas prices also rose in the aftermath of the second oil price rise. Therefore, there was less incentive for households to switch fuels at this time. Moreover, as a result of the high energy prices, the intensity of use of natural gas for space heating declined or levelled off in most countries. Against this background, an increase in gas consumption in the residential sector of about 4% per year in the period 1980–86 must be characterized as rather strong.

2.2.2 Evolution of the housing stock using natural gas for space heating

The growth in demand for space-heating gas has primarily been due to the large increase in the number of homes using natural gas for space heating. This was due to fuel switching in existing homes, but more importantly, to the installation of natural gas heating systems in new homes.

Fuel switching occurs when a heating system is replaced by one that uses a different fuel. Normally, fuel switching does not occur unless there is a considerable shift in relative prices, and usually only at the end of the system's lifetime. The 1973 oil price increases were sufficient enough to trigger fuel switching from oil to natural gas systems. Natural gas systems were more often chosen over other fuels, e.g. electricity, to replace oil systems since the necessary duct work was already in place.

The share of homes using gas heating systems increased substantially in

The evolution of natural gas demand in Western Europe

all the major West European countries from 1972 to 1985 (Fig. 2.2). In each of the countries a large portion of this growth came from the substitution of coal with gas systems when the system needed replacement. Some growth also came from the conversion from oil to natural gas systems.

The choice of space-heating systems based on natural gas in new dwellings has been important determinant of natural gas demand. In each country, the share of new dwellings using natural gas has exceeded the share of total dwellings using natural gas. The installation of gas heating systems in new homes increased sharply in the UK and West Germany (Fig. 2.3). In West Germany, the share of new homes installed with gas systems reached 55% in 1984, from only 20% in 1973. In France, where electricity is strongly promoted, the share grew less; it was 23% in 1973 and 33% in 1984. As old dwellings are replaced in the future, the share of dwellings using natural gas for space heating will comprise a larger share of total dwelling stock.

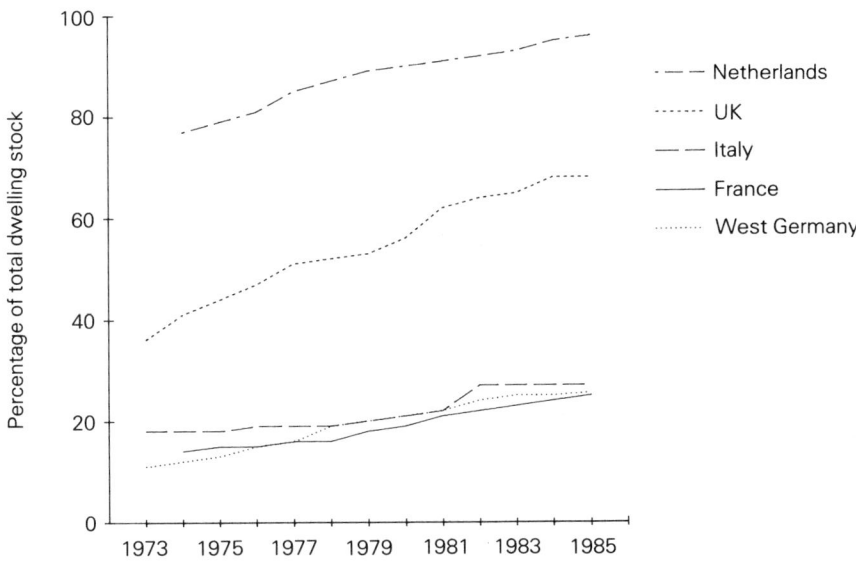

Fig. 2.2 Dwellings heated with natural gas.

2.2.3 The intensity of use

From 1973 to 1978, the intensity of use increased in every country. After 1978, the level of intensity either declined or remained constant. There are three factors that can influence the intensity of use: the characteristics of the dwelling stock (share of single-family dwellings, age and thermal integrity of

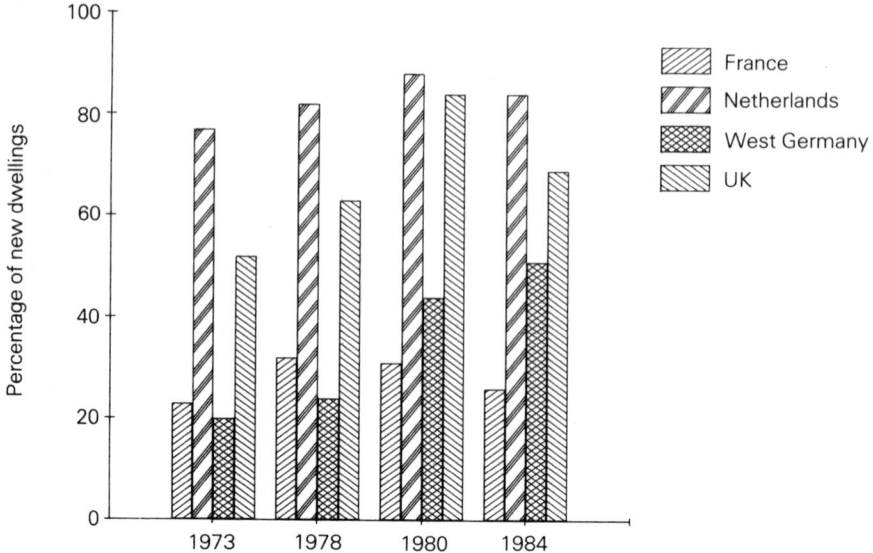

Fig. 2.3 New dwellings using natural gas for space heating.

the dwelling, and dwelling size), the characteristics of the heating equipment (share of dwellings with central heating, efficiency of the heating equipment) and the behaviour and composition of households.

The increase in the dwelling size has tended to push the level of intensity upwards. Since 1970, the average dwelling area has increased in every country except in the Netherlands.

Single-family dwellings tend to have higher intensities than multi-family homes since there are more exposed walls, and typically, there is more area to heat. Both single- and multi-family dwellings have increased in size. In France and West Germany, natural gas is used predominantly in multi-family homes because of its availability. In the Netherlands, where the gas network is quite extensive, the majority of natural gas is used in single-family homes.

Newer homes tend to have a higher level of thermal integrity than older homes because of the emergence of stricter building codes. On the other hand, they tend to be larger than older homes. These two factors offset each other with a net effect of reducing intensity. In France and West Germany, the share of new homes represents over 21% of the total stock of gas-heated homes, while in the UK the share is only 11%.

Changes in the characteristics of heating systems have caused changes in the level of the intensity of use. Two of the most important characteristics are increases in the numbers of central heating systems, and improvements in the efficiency of new gas systems. The emergence of central heating systems has tended to increase the level of intensity since central systems heat the entire home, as opposed to non-central systems which provide the flexibility to heat only one room. In the early 1960s, the majority of homes using natural gas had non-central heating systems. At present more than 60% of homes have central heating systems. In France, this share is near 90%.

Price increases that occurred during the late 1970s induced households to lower their indoor temperature. This has led to a reduction in the level of intensities in all countries – partially explaining the decline after 1978. The UK has always had the lowest average indoor temperature in the region (16 °C).

2.3 THE THEORETICAL FUEL CHOICE MODEL

The main characteristics of the discrete–continuous energy choice model are the following:

1. Before the space-heating fuel choice investment decision is taken, there are four different fuel choices available to the household. These fuels are: natural gas, fuel oil, solid fuel and electricity.[1] The fuel choice is represented by a comparison between the indirect utility functions attached to the various fuel choices where the arguments of these functions are independent variables such as fuel and equipment prices and household disposable income.
2. In new dwellings, the investment decision is assumed to follow from a comparison of indirect utility functions related to the different discrete fuel alternatives. Due to unobservable heterogeneity, these discrete choices are represented by a set of unconditional choice probabilities for the various fuels. Given certain assumptions regarding the stochastic part of the utility structure (the extreme value distribution), a traditional mulit-nomial logit choice (MNL) model is then obtained.
3. In existing dwellings, the model allows for transitions (fuel switching) among fuels. Investment decisions for space-heating fuel choice are in principle reversible at a specified cost, as opposed to a priori assumptions made in several other studies of energy demand (Goett and McFadden, 1982; Ruderman, Levine and McMahon, 1984). Changes in space-heating equipment are restricted by conversion costs, and 'habit persistence'. This makes the choice model dynamic.
4. Given the fuel choice, the intensity of use is determined as a traditional continuous choice, and the demand equation is derived from the indirect utility function by Roy's identity.

---

[1] We have not included district heating as a fuel choice.

2.3.1 The fuel choice decision in new dwellings

The investment decision in new dwellings is described as a static discrete choice. As opposed to the traditional econometric approach to energy demand modelling, discrete choice models build explicitly on the assumption that households choose among a limited number of space-heating fuels. Due to unobserved heterogeneity affecting preferences, this behaviour is represented by a set of choice probabilities, derived from the specific utility structure. For each fuel choice, we define a conditional indirect utility function, V_h, of the following type:

$$V_h = v(\mathbf{z}_h) + \varepsilon_h = v_h + \varepsilon_h \qquad (2.1)$$

where V_h is the maximum utility attainable from space heating, given that fuel choice h is chosen, $\mathbf{z}_h$ a vector of independent variables that characterize alternative h, such as prices, the user cost of capital, the level of saturation of the gas network, and other non-alternative specific variables, e.g. disposable income, v_h the structural part of the utility function, while ε_h is the stochastic term that captures the unobservable factors affecting the fuel choice. For a given fuel choice h, ε_h is assumed to vary among households.

The probability for choosing fuel h is denoted P_h, which is defined by

$$P_h = \Pr(V_h = \max_k V_k) = \Pr(v(\mathbf{z}_h) - v(\mathbf{z}_k) > \varepsilon_k - \varepsilon_h, \text{ all } k) \qquad (2.2)$$

In Dagsvik et al. (1987), a justification is given for the error terms to be independently extreme value distributed. Given this condition, $\varepsilon_k - \varepsilon_h$ is logistic distributed, and P_h can be written

$$P_h = \frac{e^{v_h}}{\Sigma_k e^{v_k}} \qquad h, k = 1, 2, 3, 4 \qquad (2.3)$$

This is the well-known MNL model.

2.3.2 Transitions among fuels

In most of the previous studies of energy demand utilizing the discrete choice approach, investment decisions are assumed to be irreversible. A consequence of this is the restriction on the investment decisions for new dwellings. Irreversibility is contradicted by the observation that fuel switching actually occurs, as discussed in section 2.2. In order to capture the fuel switching aspect, we have specified a dynamic discrete choice model which we then apply for existing dwellings. Analogous to Equation 2.2, we now define a set of conditional choice probabilities, $P_{sh}(t-1, t)$, i.e.

$$P_{sh}(t-1, t) = \Pr(V_h(t) = \max_k V_k(t) | V_s(t-1) = \max_k V_k(t-1)) \qquad (2.4)$$

where $P_{sh}(t-1, t)$ is the conditional probability that fuel h is chosen at time t, given that fuel s has maximized utility at time $t - 1$. When these transition probabilities are known, and given that the 'history' prior to time $t - 1$ does not influence the decisions, the evolution of energy choices in existing dwellings over time may be described as a Markov process.

In Dagsvik et al. (1987), the assumptions regarding the unobservable elements of the specified model guarantee that the stochastic process is a Markov process.[2] Furthermore, these stochastic assumptions introduce an important dynamic element into the model, namely autocorrelation among the fuel choices in different time periods. We have defined this as 'habit persistence', and it is expressed by the parameter α in Equations 2.5 and 2.6. When α approaches zero, the utilities of the fuel choices become perfectly correlated, while the correlation becomes negligible when α approaches infinity.

If there are no observable costs of conversion among fuels, it is shown in Dagsvik (1983) that the indicated stochastic specification implies transition probabilities of the following form:

$$P_{sh}(t - 1, t) = P_h(t) - P_h(t - 1)e^{-\alpha}, \qquad s \neq h, \qquad \alpha > 0 \quad (2.5)$$

$$P_{ss}(t - 1, t) = P_s(t) - P_s(t - 1)e^{-\alpha} + e^{-\alpha} \quad (2.6)$$

In Equations 2.5 and 2.6 the $P_h(t)$ function is the MNL probability given by Equation 2.3. It can be observed that when autocorrelation vanishes, i.e. when α is infinitely large, the transition probabilities collapse into state probabilities. This means there is no correlation among unobserved factors influencing the households' conversion decision. On the other hand, when α approaches zero, perfect autocorrelation prevails, so if the independent variables remain unchanged, households will not convert to a different fuel system.

When conversion costs, i.e. the measurable costs of changing a fuel system (difference in system costs, changes in duct work, etc.) are considered, the expressions for the transition probabilities change slightly; the terms on the right-hand side of Equations 2.5 and 2.6 are no longer identical to the MNL probabilities. This has important implications for the procedure used to estimate the model (Bartlett et al., 1987), but these difficulties are not discussed in this chapter.

2.3.3 Short-run capacity utilization

Given the fuel choice h, the energy demand is derived by applying Roy's identity. This states:

$$x_h = \frac{\partial v_h / \partial p_h}{\partial v_h / \partial y} \quad (2.7)$$

---

[2] An extremal process is specified for the utility process (Tiago de Oliveira, 1968).

where x_h is the intensity of use of fuel h. From the specified form of the indirect utility function chosen (Bartlett et al., 1987), one can obtain the following short-run demand equation:

$$\log x_h = \alpha_0 - \alpha_1 \log p_h + \alpha_3 y - \sum_{k=1}^{k=6} \beta_k z_k \qquad (2.8)$$

where p_h is the real price of fuel h^3, y the real disposable income per household, z_1 the gas network saturation level, z_2 the district heating network saturation level, z_3 the share of single-family dwellings, z_4 the average dwelling area, z_5 the number of heating degree-days, and z_6 the share of centrally heating dwellings.

2.3.4 Data and estimation results

The data utilized in this chapter are collected and maintained by the International Energy Studies Group at Lawrence Berkeley Laboratory (IES-LBL) (see Ketoff et al. 1987a,b). Combined cross-sectional and time-series data are used. It should be noted that the data are not strictly 'individual'. To be consistent with the discrete choice framework, the data may be regarded as 'grouped' data (Maddala, 1983). We utilize country averages for the 'energy'-related variables. One limitation of the IES-LBL data is that they do not provide sufficient information on conversions. If such data had been available, the estimation procedure would have been simplified. This would have allowed us to employ directly the relations for the transition probabilities used to calculate the coefficients of the model.

The estimation of the complete model is described in Bartlett et al. (1987). Recently the model has been re-estimated, and the results evaluated. Some of these results are:

1. The estimate of the energy price elasticity is -0.3. As emphasized above, this should be interpreted as a short-term elasticity.
2. The estimate of the income parameter, α_3, implies an income elasticity between 0.6 and 1.2 (it varies among countries and over time). It is interesting to compare these with estimates on the income effects for natural gas demand in Europe obtained in other recent econometric studies (Gjelsvik, Olsen and Vatne, 1987). Typically, in these studies, the income elasticity is overestimated because of the rapid evolution of the natural gas market in Western Europe. A preferable feature of our discrete–continuous choice model is that the income effect is identified in households that have already chosen natural gas for their space-heating needs. Therefore, the estimate of the income effect is not influenced by the variation in the number of gas customers during the estimation period.

[3] All energy prices are disposable income data have been converted to 1981 $US.

Residential natural gas space heating demand projections 39

3. The climate variable z_5, the share of single-family dwellings z_3, and the share of central heating z_6, were the most significant, and reasonable in magnitude of the structural variables.
4. The estimate of the correlation parameter, α, was rather low, so the model involves a relatively high level of positive autocorrelation among the elements influencing the household's fuel choice, i.e. a high degree of habit persistence. This implies that with 'smooth' developments in prices, income and other independent variables, the transition probabilities are also low.

2.4 RESIDENTIAL NATURAL GAS SPACE HEATING DEMAND PROJECTIONS

Obviously, there is a great deal of uncertainty in predicting the level of the future space-heating demand in Western Europe. First, relations estimated with historical data may be unstable. Secondly, one has to make assessments for the exogenous variables. Finally, the important elements influencing the future development of the natural gas market are not explicitly present in the model framework, e.g. supply-side factors, policy questions and environmental problems. It is important to stress that the present framework is a tool for making projections, and its strength is to track the most important impacts and relations in the market in an effective and consistent manner. The simulations presented below should primarily be regarded as demonstrations of the basic properties of the space-heating model.

2.4.1 The reference scenario

The time interval for the simulations is the period from 1984 to 2000. Focus has been placed on medium- and long-term developments, but we also project from 1984 to the present. From 1984 to the present, we use observed values for the exogenous variables for each country. In general, we do not distinguish between paths of the exogenous variable of the individual countries. The exception to the above is an assumption regarding the development of the electricity price in France, where it is assumed to decline at a rate of 1% annually from 1987 to 1995 (due to the promotion of existing nuclear power programmes). In other countries, the real price of electricity is assumed to increase by 1% per year. In addition, the simulations are based on the following assumptions:

1. After the sharp fall in the 1985/86 fuel oil prices, the oil price is assumed to follow the observed path until 1988, to stay constant until 1990, and to increase by 2% annually thereafter. The drop in the price of oil from 1985/86 until the present is assumed to have a distributed

lag influence on behaviour, with 20% of the price change occurring in the first year, 50% in the next and 30% of the price drop in the third year.
2. The price of natural gas follows the development of the price of oil, but lagged by one year.
3. Coal prices (which we use to represent the price of all solid fuels used), are kept constant until 2000.
4. The structural and independent variables included in the z-vector are assumed to be unchanged in the reference scenario, while impact calculations on some of these variables are presented below.
5. The number of new dwellings is assumed to increase by 1% annually. The assumed replacement rate is 13% per year. (This is sufficient to keep the total dwelling stock constant, or slightly increasing.)

Except for the abrupt movements in oil and gas prices during the first 3 years, the reference scenario is based on moderate changes in the variables affecting the behaviour at the various levels of the model. One motive for making these assumptions is to concentrate on a main feature of the specified model, namely the difference among the fuel shares in the new and existing dwelling stock. Since the model also tracks the evolution of the total dwelling stock, it includes very important 'structural' and 'demographic' elements. Over a long period of time a significant number of new dwellings will have entered into the market, while at the same time, some of the older dwellings will be retired from the total dwelling stock. Therefore, in the long-run, fuel choice decisions in new dwellings become important for the development of the average shares of all dwellings. This component is more important if the incentives to undertake conversions are limited. In section 2.2, it was observed that the share of natural gas used in new homes exceeded the share of natural gas in the total dwelling stock in the early 1980s. This relationship has important implications for the model simulations, demonstrated in Figs 2.4 and 2.5, illustrating the resulting fuel shares for the reference scenario.

The most striking feature of Fig. 2.4 is that it projects an increase in the share of new dwellings using fuel oil systems until after 1990. This is a direct effect of the fall in the oil price in the mid-1980s, which in the model is assumed to extend partially to consumers' expectations of future prices (the assumed expected price decreases by 25% from 1985 to 1987). Obviously, the implicit assumption regarding households' expectations may be questioned, but this problem is not addressed here. Through the 1990s, the share of new dwellings using oil is reduced again, while the share of new dwellings using electricity increases. This is most noticeable in France because of the assumptions regarding the price of electricity.

The share of total dwellings using natural gas increases steadily over the simulation period (cf. Fig. 2.5). This reflects the mechanism mentioned above; it is a direct effect of the relatively higher gas share in new dwellings. The share of electrically heated dwellings also grows, at the expense of the decline in the share of dwellings using oil and solid fuels.

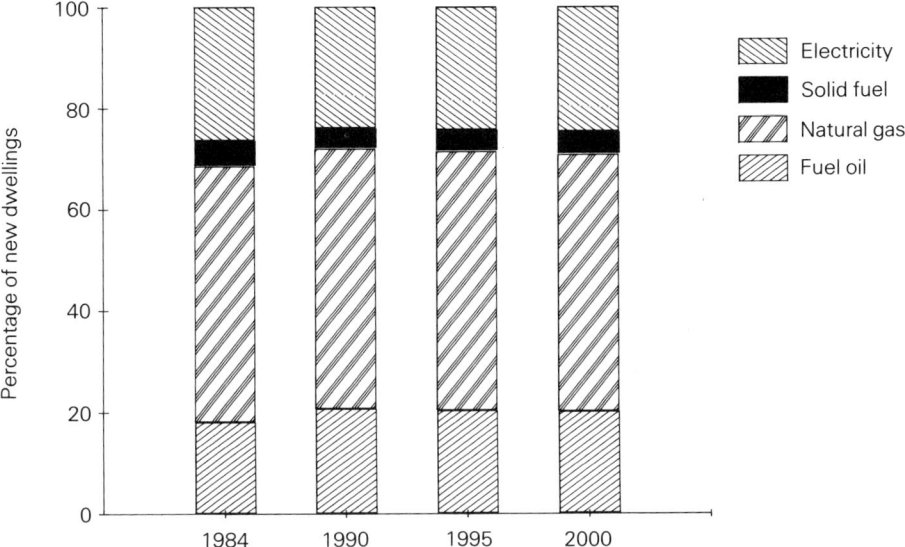

Fig. 2.4 Projected fuel shares – new dwellings.

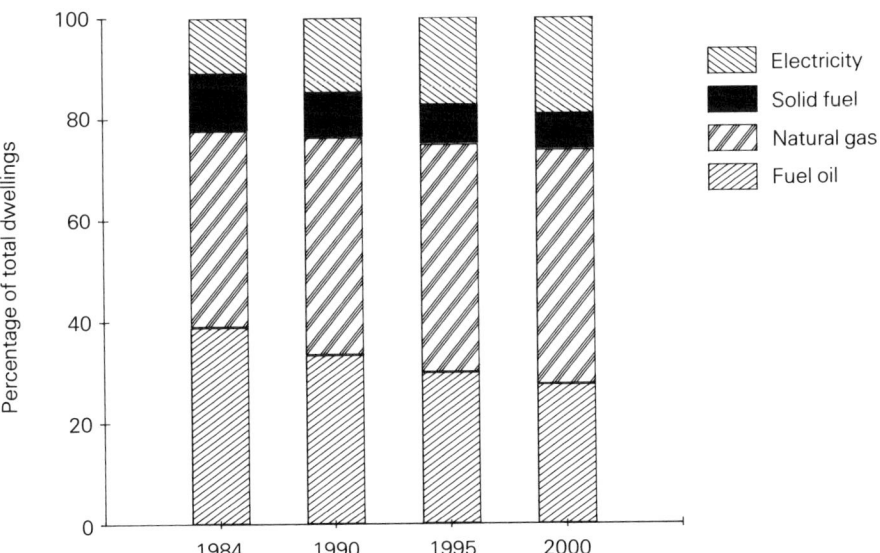

Fig. 2.5 Projected fuel shares – total dwellings.

The significance of the initially higher probability of choosing natural gas in new dwellings may be further emphasized by examining Fig. 2.6. This shows the average share of dwellings using natural gas approaching the share of new dwellings using natural gas towards the year 2000.

Figure 2.7 shows the development of the share of total dwellings using natural gas for space heating in five of the largest natural gas consuming countries in Western Europe. (In the Netherlands, natural gas already dominates the residential sector.) The highest growth is projected for West Germany, where the average share of dwellings using natural gas increases from 28% in 1984 to 45% in 2000. In the other three countries, the growth is more moderate, due to smaller deviations between the share of natural gas used in new and existing dwellings.

The part of the model describing the short-term energy-using behaviour is predominately static with the exception of the lagged response to the sharp decline in oil prices from 1985–86 until 1988. The marked fall in energy prices in 1985–86 implies an projected increase in the unit consumption. From 1988 to the end of the simulation period, the unit consumption in all countries increases slightly, in coherence with the smooth development in exogenous variables.

In addition to the development of the fuel shares, and the unit consumption, the total consumption of natural gas is dependent on the growth in the dwelling stock. The time path of gas demand as calculated by the present model simulations is shown in Fig. 2.8. For the five countries aggregated, the model projects an increase in residential consumption of natural gas for

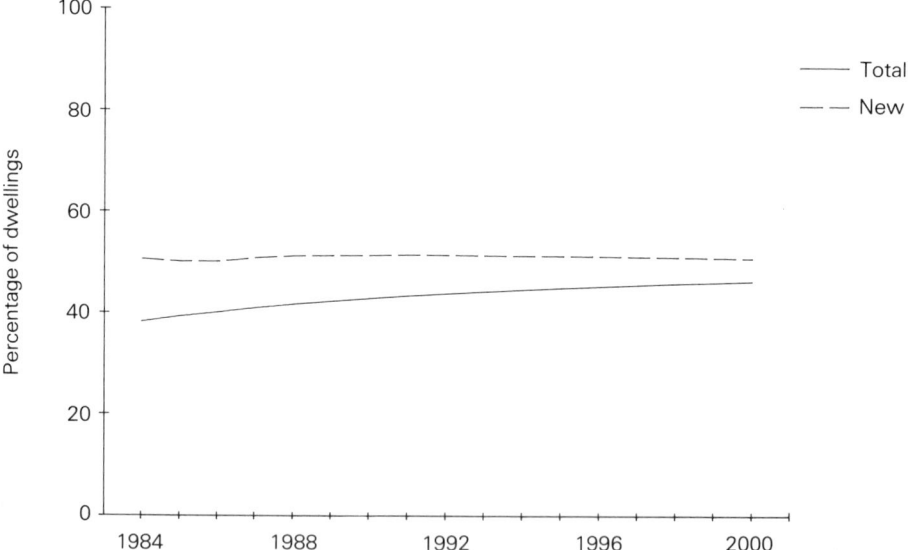

Fig. 2.6 Projected shares of total and new dwellings using natural gas.

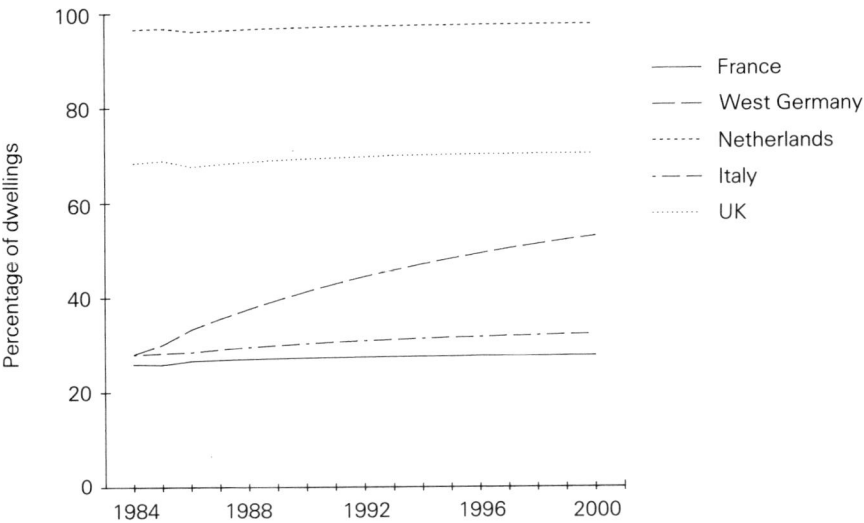

Fig. 2.7 The projected share of total dwellings using natural gas.

space heating of 38% in the period from 1984 to 2000. From a gas exporter's point of view, this may be characterized as a rather favourable prediction. In evaluating the results, it is important to remember that these results were due not only to growth in disposable income as in other models (see e.g. Gjelsvik, Olsen and Vatne, 1987). The driving force of natural gas demand in our residential demand model is the increase in new natural gas customers, particularly through investment decisions in new dwellings.

Table 2.2 shows the annual average growth rates for the various fuel shares from 1984 to 1988, and 1988 to 2000. In the latter period, total energy demand is projected to remain constant. The projected shares of natural gas and electricity used to meet total demand increase, while the projected shares of oil and solid fuels decrease. The growth in natural gas demand is due to a slight increase in the projected unit consumption, and to the increases in the total dwelling stock driven by gas investments in new dwellings. In the 1990s, we project the share of new dwellings using natural gas for space heating will decline somewhat, while the share of new dwellings using electricity or solid fuels for space heating will rise. The increase in the share of new dwellings using electricity for space heating is most noticeable in France.

Implicit in these calculations it is assumed that new households will have access to the local natural gas grid. This variable is not treated explicitly in the formal model, even though 'gas network saturation' is represented as an independent variable, z_1. This is defined as the share of dwellings located in a 'gas area'. However, this does not necessarily imply that natural gas is

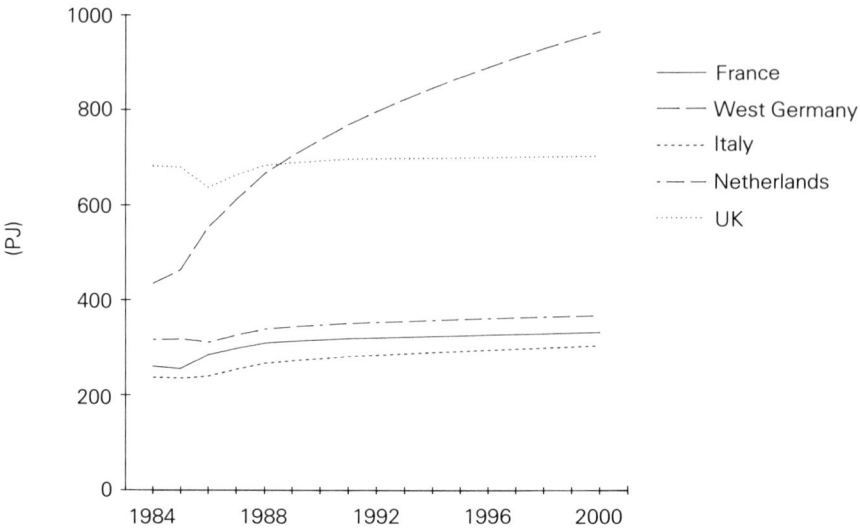

Fig. 2.8 Projected natural gas consumption for space heating.

Table 2.2 Residential energy demand in Western Europe. Growth projections 1984–2000 (annual average growth rates of fuel shares, %)

	1984–88	1988–2000
New dwellings		
Oil	3.6	−0.3
Gas	0.3	−0.1
Solid fuel	−5.4	0.8
Electricity	−2.4	0.3
Total dwellings		
Oil	−2.5	−2.0
Gas	1.9	0.9
Solid fuel	−4.1	−2.6
Electricity	5.3	2.9
Unit consumption		
Oil	2.8	−0.0
Gas	2.3	0.5
Solid fuel	−0.5	−0.7
Electricity	−0.0	0.4
Total consumption		
Oil	0.2	+1.9
Gas	4.2	1.4
Solid fuel	−4.7	−3.1
Electricity	5.0	3.3
Total	1.4	−0.0

Residential natural gas space heating demand projections

actually available for the individual household. In order for the natural gas market to grow at the projected rate, it would probably require special efforts by both public authorities, and special action taken by or towards the gas distribution companies. Still, we believe that the features of the model demonstrated by the present simulations point to some very important aspects for the future evolution of the natural gas market. To deny the possibility of future growth implicitly means assuming the share of new dwellings choosing natural gas for heating would have to be significantly reduced in the future.

2.4.2 Impact calculations

In order to demonstrate further some empirical features of the model, and to stress the impact of some independent variables on future natural gas demand, we have made some additional calculations. We have focused on the following three variables:

1. the time path for the price of natural gas, p_1;
2. the expansion of the gas distribution network, z_1;
3. the penetration of central heating systems in the dwelling stock, z_6.

For these three variables we have made changes in the growth paths compared to the reference scenario, and we have rerun the model for each alternative. In scenario (1), the price of natural gas is assumed to differ from the price path of fuel oil. More precisely, the real natural gas price is assumed to be constant over the entire period (20% lower in the year 2000 than in the reference scenario). This is an important factor concerning the future competition of natural gas and other fuels. The underpinning for this assumption is the environmental aspect (where natural gas may be favoured over fossil fuel because of taxes levied on fuels other than natural gas), and also the possible removal of monopoly power in the European markets. For illustrative purposes, comparison (1) is presented for West Germany.

Another issue considered is the future evolution of the natural gas network. Again, West Germany is used as an example. In this simulation, z_1, i.e. the share of potential natural gas customers, is assumed to increase by 1% annually until the year 2000.

When studying the effect of increasing the share of centrally heated dwellings, (case 3), the UK is taken as an example, where there is believed to be a potential for future increases in the penetration of such systems. On an *ad hoc* basis, the share is increased from 54% in 1984 to 68% in 2000.

The results from the impact calculations are summarized in Table 2.3 for the central variables of the model. When the price of natural gas is kept constant, and oil and electricity prices gradually rise, the tendency to choose natural gas for space heating is increased. The effect is strongest in new dwellings (56%), and somewhat less (3.4%) in the total dwelling stock. The

46 Residential energy demand

Table 2.3 Alternative simulations. Deviations from the reference scenario, the year 2000 (%)

	Gas share, new dwellings	Gas share, total dwellings	Unit consumption	Total consumption
(1) p_1 (W. Germany) (20% decrease)	5.6	3.4	6.8	10.2
(2) z_1 (W. Germany) (0.75–0.88)	5.4	3.7	0	3.7
(3) z_6 (UK) (0.54–0.68)	0	0	6.6	6.6

latter is influenced by the former – as new dwellings are added to the total dwelling stock. In additon, the more favourable price position of natural gas relative to other fuels implies an increase in the conversion rates. The impact on the intensity of use reflects the short-term price elasticity reported earlier.

The assumed expansion of the West German distribution network has a significant effect on natural gas consumption (and the impacts on the natural gas shares are coincidentally similar to alternative 1). When evaluating this result, the definition of the concept of network saturation should be kept in mind. Unit consumption is not affected by the expanding natural gas grid. On the other hand, unit consumption is affected when there is an increase in the penetration of central heating. In the case of the UK, the specified change implies an increase in the unit consumption, and thus, a change in total consumption of 6.6%.

2.5 SUMMARY AND CONCLUDING REMARKS

The main purpose of this chapter is to provide the reader with information regarding the main determinants of natural gas demand, and to demonstrate how this model can be used to give projections of demand for space heating in Western Europe. For a group of countries, we have simulated a reference scenario from 1984 to 2000. Except for the drop in oil prices in 1985 and 1986, and an assumed decrease in the price of electricity in France, this scenario is characterized by limited changes in relative fuel prices, moderate income growth and no changes in the structural variables.

This simulation emphasizes some of the main features of the model. The share of natural gas dwellings tends to increase over time as a result of the share of dwellings using natural gas exceeding the average share of the total dwelling stock. In the present calculations, this is the major underlying factor behind the close to 40% growth in space heating gas demand occurring between 1984 and 2000.

ACKNOWLEDGMENTS

John K. Dagsvik has contributed significantly to the specification of the empirical model used. In addition, comments from Lee Schipper and Olav Bjerkholt are appreciated.

REFERENCES

Bartlett, S., Dagsvik, J. K., Olsen, Ø. and Strøm, S. (1987) Fuel choice and the demand for natural gas in Western European households, *Discussion paper no. 23*, Central Bureau of Statistics, Oslo.

Dagsvik, J. K. (1983) Discrete dynamic choice: an extension of the choice models of Thurstone and Luce. *Journal of Mathematical Psychology*, 27, 1–43.

Dagsvik, J. K., Lorentsen, L., Olsen, Ø. and Strøm, S. (1987) Residential demand for natural gas: a dynamic discrete–continuous choice approach, in *Natural Gas Markets and Contracts* (eds. R. Golombek, M. Hoel and J. Vislie), North-Holland Publishing Company, Amsterdam.

Gjelsvik, E., Olsen, Ø, and Vatne, B. H. (1987) Utsikter for det vest-europeiske gassmarkedet: Noen beregningsresultater for gassforbruket fram til 2000. *Økonomiske analyser*, no. 87/8. Central Bureau of Statistics, Oslo (in Norwegian).

Goett, A. and McFadden, D. (1982) *Residential End-use Energy Planning System (REEPS)*, Report EA-2512, Electric Power Research Institute, Palo Alto, California.

Ketoff, A., Bartlett, S., Hawk, D., Meyers, S. and Sanipper, L. (1987a) Evolution of residential energy demand in OECD countries Part one: 1970–1984 data base. LBL-222467, International Energy Studies Group, Lawrence Berkeley Laboratory: California, USA

Ketoff, A., Bartlett, S., Hawk, D. and Meyers, S. (1987b) Residential energy demand in six OECD countries: Historic trends and future directions. LBL-22642, International Energy Studies Group, Lawrence Berkeley Laboratory, California, USA

Maddala, G. S. (1983) *Limited Dependent and Qualitative Variables in Econometrics*, Cambridge University Press, New York.

Ruderman, H., Levine, M. D. and McMahon, J. E. (1984) The behaviour of the market for energy efficiency in residential appliances including heating and cooling equipment. LBL-15304, Energy Analysis Program, Lawrence Berkeley Laboratory, California, USA

Tiago de Oliveira, J. (1948) External Processes: Definitions and properties: Publ. Inst. Stat., University of Paris, XXVII, 2

3

The European gas market as a bargaining game

MICHAEL HOEL, BJART HOLTSMARK

AND JON VISLIE

3.1 THE EUROPEAN GAS MARKET

Natural gas is usually transported through pipelines from the gas fields to the users. Due to this structure of the transmission system, which involves large transportation costs, there is no world market for natural gas. Instead we find a set of segmented or geographically separated markets. In Western Europe, it is useful to distinguish between continental Europe and the UK: continental buyers are able to buy gas from several sources, whereas the UK must rely on Norway as the only source in addition to its domestic production (disregarding the possibility of LNG imports).

There are four major countries exporting natural gas to continental Europe: the Netherlands, Algeria, the USSR and Norway, and each country is linked to the Continent by a transmission system. Among the buyers, four countries, namely West Germany, France, Italy and the UK, had 73% of the European consumption of natural gas in 1987. Moreover, in each of the selling and buying countries sales and purchases are dominated by one large company, such as British Gas in the UK and Ruhrgas in West Germany. These companies are in several cases under strong regulation by the government. As the European gas market is dominated by only a few agents on each side of the market, we could characterize this special market structure as 'bilateral oligopoly'.

It is well known that even for bilateral monopoly situations, traditional market theory does not predict a unique outcome for price and quantity. Within a bilateral monopoly framework, price and quantity are usually determined through negotiations between the parties involved. Future prices for natural gas are therefore extremely difficult to predict.

The main focus of the present chapter is an analysis of bargaining when there is more than one seller and/or one buyer of natural gas. In section 3.2 we discuss some aspects which are important for the outcome of negotiations between a seller and a buyer of natural gas. Section 3.3 gives a theoretical analysis of a simple case of one buyer and two sellers. The core of the game is derived and discussed.

In section 3.4 the simple model of section 3.3 is extended to a numerical model of a four-player bargaining game. The four players are the USSR and Norway as gas sellers and continental Europe and the UK as gas buyers. The volumes and prices of gas for the year 2010 belonging to the core are given in section 3.5, both for a reference case and for several alternatives.

3.2 BARGAINING BETWEEN TWO PARTIES

The simplest bargaining situation between two parties is one in which the traded quantity is fixed (provided an agreement is reached). In this case the negotiated price must lie somewhere between the reservation prices of the two parties. The seller's reservation price is the lowest price which is compatible with the seller being at least as well off with an agreement as without. Similarly, the buyer's reservation price is the highest price which is compatible with the buyer being at least as well off with or without an agreement.

It is, however, not reasonable to assume that the gas quantity is fixed before negotiations start. The gas quantity is usually determined simultaneously with the gas price through the negotiations. In order to understand the relationship between the price and quantity of natural gas, one must examine the character of the buyer in more detail.

In negotiations of gas trade, the buyer is a distribution company (or a consortium of several distribution companies) for natural gas. This distribution company buys gas from the gas seller and resells it to the ultimate users (households, industry, etc.). Provided that the ultimate users are not rationed, there is a unique relationship between the price these users are willing to pay and the gas quantity they use (given other energy prices, income, etc.). In spite of this unique relationship, there need not be a corresponding unique relationship between the price and quantity in a negotiated agreement between a gas seller and the gas distribution company buying gas. There are two reasons for this. In the first place, the gas distribution company usually buys gas from several sources, so that a relationship between average price and total quantity does not imply a particular relationship between price and quantity in each gas deal. In the second place, even though the quantity of gas sold by the distribution company determines the price paid by the ultimate users, and therefore this company's total revenue, this does not determine the price paid by the gas distribution company for the gas it buys and resells. The lower the gas price paid by a gas distribution company, the higher will the profits of this company be (for a given gas quantity). Whatever quantity a seller and a distribution company agree upon, the distribution company will therefore try get as low as possible a price in the negotiations.

To see how gas prices and quantities may be determined in negotiations, consider the simple case in which quantities from all other gas sellers are given. Moreover, to simplify we ignore distribution costs and price discrimination between different gas users. With these simplifications we get a unique relationship between the quantity of gas delivered by the gas seller under consideration and the ultimate gas users. This relationship is downward sloping, as in Fig. 3.1, since it represents the demand curve of the ultimate gas users.

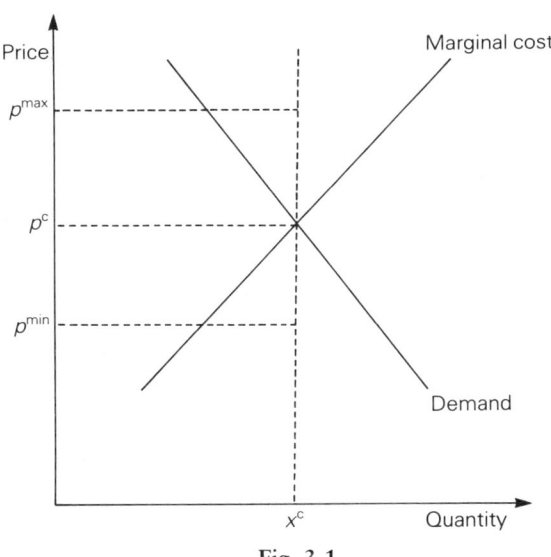

Fig. 3.1

The upward-sloping curve in Fig. 3.1 represents the marginal cost of extracting and transporting natural gas.[1] The intersection between the two curves represents the competitive equilibrium, giving a gas quantity x^c and price p^c.

When there is complete information, it is usually assumed in bargaining theory that the two parties reach an agreement which is Pareto-optimal for them. In the present context, this implies that the gas quantity is chosen so that the total gains to the two parties are maximized, provided the price can be negotiated independently of which quantity is agreed upon. The quantity

---

[1] Implicitly, this figure assumes that any economies of scale in the transmission sector are dominated by diseconomies of scale in the extraction sector.

which maximizes the total gains to the parties depends on the objective of the gas distribution company. Assume first that the distribution company is a public utility which represents the interest of all citizens of the country. Ignoring income distribution issues (i.e. assuming that lump-sum taxation is possible), a reasonable objective of the distribution company is to maximize the sum of the company's profit and the consumer surplus. The gas seller wants his profit (i.e. revenue minus costs) to be as large as possible. In this case the total gain to the two parties is equal to the area under the demand curve minus total extraction and transportation costs. It is well known that the competitive quantity x^c maximizes this total gain.

Given the quantity x^c, the buyer wants the price to be as low as possible, while the seller wants the price to be as high as possible. The outcome of the negotiation will be somewhere between the two reservation prices p^{min} and p^{max}. Here p^{min} is defined by

$$p^{min}x^c - c(x^c) = 0$$

where $c(x)$ is the total cost function; p^{max} is defined by

$$u(x^c) - p^{max}x^c = 0$$

where

$$u(x^c) = \int_0^{x^c} (p(x)dx)$$

where $p(x)$ is the inverse of the demand curve.

Notice that the ultimate consumers pay p^c for the quantity x^c no matter what the negotiated price between the seller and the distribution company is. In particular, if the negotiated price is above p^c, the distribution company makes a financial loss. The reason why it nevertheless may accept such an outcome is that the consumer surplus will exceed the company's financial loss as long as the negotiated price is below p^{max}.

Consider next the case in which the distribution company is a profit-maximizing firm. In this case the total gains to the two parties is simply the difference between total revenue from selling to the ultimate users minus the total extraction and transportation costs of the gas volume. The gas quantity which maximizes these total gains is the monopoly quantity x^M in Fig. 3.2. The gas consumers pay p^M for this gas quantity. Since the distribution company in this case will not accept a deal giving a financial loss, p^M is also the reservation price of the distribution company.[2] The reservation price for the seller is in this case $^*p^{min}$ (defined by $^*p^{min}x^M - c(x^M) = 0$). The negotiated price will therefore be somewhere in the range between $^*p^{min}$ and p^M.

From the discussion above it is clear that both the gas quantity and the interval of possible negotiation outcomes for the gas price depend on the

---

[2] With positive distribution costs, the reservation price is of course below p^M.

nature of the company which buys gas from the producer. In practice, natural gas distribution companies are generally either public utilities or regulated private firms. However, this does not imply that the case illustrated in Fig. 3.2 is irrelevant. Even publicly owned or regulated companies may be strongly influenced by self-interest in addition to their role of representing the citizens of the country.

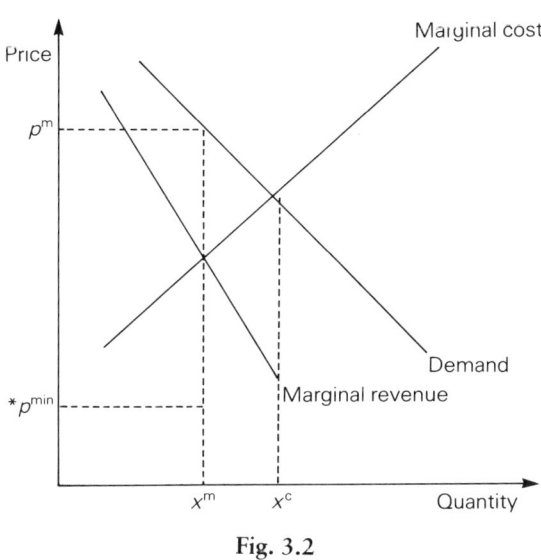

Fig. 3.2

So far, nothing has been said about where in the possible range $[p^{min}, p^{max}]$ or $[^*p^{min}, p^M]$ the negotiated price will be. This question has been treated, for example, by Hoel and Vislie (1987) and Vislie (1987), who also explicitly consider the dynamics of the bargaining problem due to the exhaustible nature of natural gas. In the present chapter, we shall instead concentrate on extending the bargaining situation to one in which there are more than two parties involved.

3.3 THE CORE OF A SIMPLE THREE-PLAYER BARGAINING GAME

Negotiations about natural gas deliveries are usually carried out on a bilateral basis. The existence of several other potential gas buyers and sellers is nevertheless important for the bargaining outcome. In the preceding section we argued that the bargaining outcome would lie between the reservation prices of the buyer and the seller. A similar requirement applies when there is more than two parties: also in this case each seller (buyer) only accepts a gas deal if this gives him a price which is not lower (higher) than his reservation price. Moreover, it is reasonable to require that each subgroup of players (with two or more members) should be at least as well off from a set of gas deals as this subgroup could be without having any gas trade with the rest of the participants in the gas market. A set of deals which satisfies these requirements belongs to the core of the multi-player bargaining game.

To illustrate the concept of the core, we shall consider a simple three-player game. We have two sellers, players 1 and 2, and one buyer, player 3. Each seller wishes to sell one unit of gas, and their reservation prices are c_1 and c_2, with $0 < c_1 \leq c_2$. The buyer wishes to buy one unit of gas and has a reservation price b, where $b > c_2$ is assumed.

In the absence of seller no. 2 (1), we would have a situation similar to the situation discussed in section 3.2. One unit of gas would be sold from seller no. 1 (2) to the buyer, and the negotiated price would be in the range $[c_1,b]$ ($[c_2,b]$).

With two (potential) sellers, the situation is more complex. Denote the pay-off (or benefit) to the three players of a set of deals by π_1, π_2 and π_3. Gas deals in the core must satisfy

$$\pi_1 + \pi_2 + \pi_3 = b - c_1 \tag{3.1}$$

$$\pi_i \geq 0 \quad i = 1, 2, 3 \tag{3.2}$$

$$\pi_1 + \pi_2 \geq 0$$
$$\pi_1 + \pi_3 \geq b - c_1 \tag{3.3}$$
$$\pi_2 + \pi_3 \geq b - c_2$$

Equation 3.1 is a requirement that the solution is Pareto-optimal for the three players. In other words, the players will not settle with a set of deals when an alternative set of deals can improve the situation for all players. This requirement implies that the buyer gets one unit of gas, and that this unit of gas is delivered by the seller with the lowest reservation price. (If the sellers have equal reservation prices the gas may be delivered by either of the two sellers.)

The inequalities 3.2 are requirements of individual rationality, i.e. that none of the three players can improve their situation by staying out of the natural gas market.

The inequalities 3.3 are requirements of group rationality, and are in a sense a generalization of Equation 3.2. Since the two sellers cannot achieve anything without the buyer, we have the trivial requirement $\pi_1 + \pi_2 \geq 0$ (which also follows from Equation 3.2). Seller no. 1 and the buyer can achieve $b - c_1$ without the participation of seller no. 2. These two players therefore only accept deals which satisfy $\pi_1 + \pi_3 \geq b - c_1$. The same argument explains the last inequality in Equation 3.3.

Straightforward manipulation of Equation 3.1–3.3 gives

$$\pi_2 = 0$$
$$0 \leq \pi_1 \leq c_2 - c_1 \qquad (3.4)$$
$$b - c_2 \leq \pi_3 \leq b - c_1$$

Consider first the case in which the two sellers have equal reservation prices, i.e. $c_1 = c_2 = c$. From Equation 3.4 this gives $\pi_1 = \pi_2 = 0$ and $\pi_3 = b - c$. In this case the gas is sold at a price equal to the common reservation price of the sellers. The whole gain from the trade therefore goes to the buyer. In this case the buyer can thus 'set the sellers up against each other', and thereby capture the whole gain himself.

If $c_1 < c_2$, we still get $\pi_2 = 0$, which is consistent with the fact that seller no. 1 now supplies the gas to the buyer. In this case seller no. 1 might get some of the gains from the trade, but this pay-off cannot exceed $c_2 - c_1$. Denoting the gas price by p, we see that $\pi_1 \equiv p - c_1 \leq c_2 - c_1$ implies $p \leq c_2$. The buyer's pay-off cannot exceed $b - c_1$, i.e. $p \geq c_1$ (from $\pi_3 \equiv b - p \leq b - c_1$). We thus see that the negotiated gas price must lie in the range $[c_1, c_2]$.

In this example, it is only seller no. 1 and the buyer who are involved in any gas trade (for $c_1 < c_2$). The existence of player no. 2 as a potential seller is, however, important for the possible outcomes of the negotiations between seller no. 1 and the buyer. In the absence of player no. 2 the negotiated price lies in the range $[c_1, b]$. Hence, the existence of player no. 2 as a potential seller narrows down the possible bargaining outcomes to $[c_1, c_2]$, which means a better bargaining position for the buyer.

We shall conclude this section with a brief discussion of which of the prices in the core are likely candidates for the outcome of the bargaining process. In the absence of seller no. 2 all prices in the range $[c_1, b]$ are in the core. For this extremely simple case most bargaining theories would suggest that the bargaining outcome is the price in the middle of these two reservation prices, i.e. that the negotiated gas price is $(c_1 + b)/2$. The well-known (symmetric) Nash bargaining solution is an example of a bargaining theory predicting this outcome. The Nash bargaining solution for this simple case has the reservation prices c_1 and b as the 'disagreement point', leading to the bargaining solution $(c_1 + b)/2$.

Assume now that $c_1 < c_2$ and that seller no. 2 offers his gas at the price c_2, i.e. at a price equal to his reservation price. In such a situation we might expect that the relevant reservation price for the buyer in his negotiations with seller no. 1 is c_2, since the buyer can always obtain his gas at the price c_2. This line of reasoning suggests that b should be substituted by c_2 in the expression for the equilibrium price, i.e. instead of $(c_1 + b)/2$ the bargaining outcome for the price should be $(c_1 + c_2)/2$. Notice that this candidate for a bargaining outcome belongs to the core, which contains all prices in the range $[c_1, c_2]$.

The use above of the outside option c_2 as the disagreement point in the Nash bargaining solution is quite common in the bargaining literature. However, recent theory of bargaining formulated as a non-cooperative extensive form game has shown that this way of using outside options is not necessarily correct (cf. e.g. Binmore, Rubinstein and Wolinsky, 1986; Binmore, 1987). For our example this theory suggests that the introduction of seller no. 2 changes the bargaining outcome for the price from $(c_1 + b)/2$ to $\min[c_2, (c_1 + b)/2]$ (and not to $(c_1 + c_2)/2$). In other words, the so-called 'outside option principle' will work here; an offer of c_2 from seller no. 2 only influences the negotiated gas price if this offer is better for the buyer than the bargaining outcome in the absence of the outside option (in which case the negotiated gas price will be equal to the outside option). Notice that

$$c_1 < \min[c_2, (c_1 + b)/2] \leq c_2$$

i.e. the proposed bargaining solution belongs to the core.

In the example above it was assumed that seller no. 2 gave a binding offer equal to c_2. In practice a buyer is often uncertain about what alternative offer he can get if he breaks the negotiations with the original seller. Even if there exists one or several alternative sellers, they will seldom give binding price offers before serious negotiations with the buyer take place. Possible outcomes of the bargaining process when these types of complications are allowed for have been analysed by, for example, Hoel (1986), Sutton (1986) and Vislie (1988). These analyses reveal that the outcome of a bargaining game is quite sensitive to details of the rules of the game. This confirms that in practice it is difficult to predict the likely outcome of negotiations.

3.4 THE CORE OF THE EUROPEAN GAS MARKET

We have made numerical calculations of the core of a somewhat extended version of the bargaining game in section 3.3. In this section we give a rough description of the model (see Hoel, Holtsmark and Vislie (1987) for further documentation). A more detailed discussion of assumptions, parameter values, etc. is given in the Appendix.

3.4.1 A four-player game

The model includes two sellers and two buyers. The sellers are the USSR and Norway, while the buyers are continental Europe and the UK. In 1987 these two buyers imported a total of 130 bcm natural gas (including imports from the Netherlands). The imports were divided between the USSR

(30%), Norway (23%), Algeria (18%), the Netherlands (27%) and others (2%). In the model, imports from Algeria and the Netherlands are given exogenously. Moreover, continental Europe is treated as one buyer. In 1987, the total imports of 118 bcm natural gas to continental Europe were divided between West Germany (38%), France (23%), Italy (20%) and others (19%).

At present, there is no pipeline across the Channel. This means that gas sales from the USSR to the UK are excluded. The relevant gas flows are therefore from the USSR to continental Europe and from Norway to continental Europe and the UK.

In our model, we may assume that this trade structure remains valid in the future. However, we can alternatively assume that a pipeline across the Channel will be built in the future. In this case gas exports from the USSR to the UK are possible. Since such gas must pass through continental Europe, exports from the USSR to the UK need the consent of continental Europe. It therefore seems most reasonable to assume that the total gains possible for the USSR and the UK without cooperation of the other two players are zero even if a pipeline across the Channel is built. In the model it is also possible to assume alternatively that transport of gas from the USSR to the UK is always possible, provided the transport costs are paid. As long as transport costs are not too high, the gains to the USSR and the UK without cooperation with other players is positive under this assumption.

3.4.2 Cost structure

The cost structure in the model is extremely simple. Extraction costs are of the 'inverse L' type, i.e. unit extraction costs are constant up to an exogenous capacity limit. Extraction costs for gas from the USSR are assumed to be somewhat lower than for gas from Norway (cf. Table 3.1).[3]

Unit transport costs depend only on where the gas is transported to and from. Transportation from the USSR to the gas-using countries costs considerably more than transportation from Norway. As is clear from Table 3.1, the sum of extraction and transportation costs are therefore slightly higher for the USSR than for Norway.

In addition to transport costs, there are country-specific distribution costs. Like other costs, these distribution costs are assumed to be proportional to the gas quantity.

3.4.3 Import diversification

From Table 3.1 we can see that gas from the USSR has higher total unit

---

[3] Costs are measured in 1987 Nkr per m^3. At the exchange rate of September 1989 (Nkr 7.25 per $US), Nkr 1 per m^3 is equal to $US4.30 per million BTU.

Table 3.1 Unit costs of production, transport and distribution of natural gas (1987 Nkr per m^3)

From:	USSR		Norway	
To:	Continental Europe	UK	Continental Europe	UK
Unit production costs	0.12	0.12	0.20	0.20
Unit transport costs	0.35	0.45	0.22	0.15
Unit distribution costs	0.55	0.66	0.55	0.66
Total unit costs	1.02	1.23	0.97	1.01

costs than gas from Norway. This is true both for gas deliveries to continental Europe and the UK (provided a pipeline across the Channel is built). Without any capacity restriction on Norwegian gas, we thus get a situation similar to the one described in section 3.2: the gas seller with the highest unit costs (USSR) will not sell any gas in the solutions belonging to the core. However, this situation does not occur in our model, since we assume that there is a binding capacity limit to Norwegian gas production (60 bcm in the reference case).

Even in the absence of any production capacity constraints, it is unreasonable to assume that an importing country buys all its gas from the seller with the lowest cost. It is more reasonable to assume that the importing countries wish to diversify their gas imports somewhat.[4] The diversification argument is included in our model in a somewhat crude way by setting an exogenous limit on imports from the USSR as a share of total gas consumption in the buying countries. In our reference case we have set this limit equal to 30% both for continental Europe and the UK. This corresponds to the limit recommended by the IEA to their member countries. This import limit implies that it would be possible for Norway to sell gas even if Norway had higher costs than the USSR, and even if there was no capacity limit on production in the USSR (see Vislie, 1989 for further discussion of this market share requirement).

[4]For a further discussion of such diversification issues, see Hoel and Strøm (1987) and Manne, Roland and Stephen (1986).

Results

3.5 RESULTS

We have used the model to see what the core of the bargaining game implies for gas quantities and prices in 2010. In our reference case we have assumed that the gas-importing companies are pure profit-maximizing firms. We have also assumed that a pipeline across the Channel will be built, but that transport of gas from the USSR to the UK needs the consent of continental Europe.

Table 3.2 gives the gas quantities in the core in 2010 for the reference case. These quantities are, of course, strongly influenced by exogenous variables and parameters, such as GNP growth, demand elasticities, the Norwegian production capacity, and the maximal share of USSR imports in total gas consumption.

Table 3.3 gives minimum and maximum prices for the sellers and buyers (excluding the constant unit costs of transport and distribution). The most striking feature of Table 3.3 is the wide range of possible prices for the players which are consistent with solutions in the core.

Table 3.2 Gas quantities in the core in 2010 (bcm)

	UK	Continental Europe	Continental Europe+UK
Imported from Norway	17	43	60
Imported from the USSR	20	53	73
Imported from Norway and the USSR	37	96	133
Imported from Algeria and the Netherlands	0	55	55
Total gas consumption	67	176	243

Table 3.3 Price ranges in the core in 2010 (1987 Nkr per m^3)

	Minimum	Maximum
Gas from Norway	0.32	1.05
Gas from the USSR	0.12	0.58
Gas to the UK	0.20	0.86
Gas to CE	0.21	0.83

Table 3.4 indicates which divisions of the total surplus (106 bn 1987 Nkr) are possible for solutions in the core. Norway and continental Europe are guaranteed part of the total surplus, while the USSR or the UK may end up getting none of the total surplus from the gas trade. In fact, both the

USSR and the UK may end up receiving zero pay-off. The reason for this is that in the reference scenario, it is assumed that transportation of gas from the USSR to the UK is only possible with the consent of continental Europe. The total gain possible for the USSR and the UK without cooperation from the other players is therefore zero.

Table 3.4 Distribution of total surplus in the core of 2010 (bn 1987 Nkr)

	Minimum	Maximum
Pay-off for Norway	7	51
Pay-off for the USSR	0	34
Pay-off for continental Europe	16	74
Pay-off for the UK	0	25
Total pay-off		106

The maximal surplus Norway can get is 1987 Nkr51 bn. With 2% yearly real GNP growth in Norway from 1987 to 2010, this corresponds to the surplus from gas exports being maximally 6% of Norway's GNP in 2010.

Finally, Table 3.5 shows how some of the relevant variables depend on which assumptions we use. Norwegian exports are in all of the cases considered to be determined by the Norwegian capacity limit. However, exports from the USSR and price range for Norway are quite strongly affected by which assumptions are made.

In the reference case it is assumed that a pipeline across the Channel will be built. Without a pipeline (scenario B), the USSR cannot export to the UK. This import loss to the UK is partly compensated by increased imports from Norway. This in turn reduces Norwegian exports to continental Europe (due to the Norwegian capacity constraint). Since the constraint on imports from the USSR is as a percentage of total imports, the reduction in imports from Norway implies reduced imports from the USSR to continental Europe. Total Soviet exports are thus considerably lower without a pipeline across the Channel than with.

Norway's bargaining position is stronger without a pipeline across the Channel, moving the price range for Norway upwards. However, the difference between the two cases is modest, at least for price and pay-off intervals for Norway.

With a lower international GNP growth, the demand for gas in 2010 will be lower. In scenario C we see that this implies lower gas exports from the USSR, and a slight reallocation of Norwegian exports from continental Europe to the UK. The price range for Norway is also somewhat lower the

Table 3.5 Exports (bcm) and prices (1987 Nkr per m^3) under alternative assumptions

	Exports from Norway		Exports from the USSR		Price for Norwegian gas	
	To Continental Europe	To UK	To Continental Europe	To UK	Min	Max
A. Reference case	43	17	53	20	0.32	1.05
B. No pipeline across Channel	33	27	48	0	0.38	1.20
C. 1.5% demand growth (reference case = 2%)	37	23	50	11	0.28	0.89
D. 25% reduction in transport costs from the USSR	43	17	53	20	0.28	1.09
E. Max. share of imports from the USSR 40% (ref.case = 30%)	27	33	71	6	0.27	0.75

lower the demand growth is. However, for this alternative the consequences for Norway are also relatively modest.

In the reference case, it is assumed that Norwegian gas has a lower unit cost of production plus transportation than gas from the USSR. Cost estimates for the USSR are, however, very uncertain. In scenario D we therefore consider a 25% reduction of transportation costs from the USSR. This makes the total unit costs of gas from the USSR to continental Europe lower than for Norwegian gas. Without any restriction on imports of gas from the USSR, this could have a dramatic impact on Norwegian gas exports. However, the restriction that gas from the USSR should not exceed 30% implies that the cost of gas from the USSR is of far less significance for Norway. As is clear from Table 3.5, exports from Norway as well as the USSR are unaffected by this reduction in Soviet costs. From Table 3.5 we also see the somewhat surprising result that reduced costs for gas from the USSR does not unambiguously move the price range for Norwegian gas downwards: the maximal price Norway can get is higher the lower the costs of USSR gas are. In other words, the aggregate gain of reduced costs might not be divided only between the USSR and the gas buyers. Some of the aggregate gain might go to Norway, through buyers paying more for Norwegian gas although their average price for gas imports is reduced.

The importance of the restriction on imports from the USSR is illustrated by scenario E, where it is assumed that maximal share of imports from the USSR is 40% instead of 30%. It is clear from Table 3.5 that this only gives a modest increase in total Soviet exports. However, much more of the exports now go to continental Europe, while a larger share of Norway's gas now is sold to the UK. The reason for this reallocation of exports is that continental Europe in a sense is more constrained by the import limit on Soviet gas than the UK. Although a 30% limit is binding for both continental Europe and the UK, a 40% limit is only binding for continental Europe (in this case only 8% of the UK's imports come from the USSR). If there was no limit on gas imports from the USSR, continental Europe would import 48% of its gas from the USSR, while the UK would only import from Norway.

It is clear from Table 3.5 that the increase in the limit of imports from the USSR gives quite a significant downward movement of the range for Norwegian prices.

3.6 CONCLUSIONS

Since natural gas is traded on a bilateral basis, via a pipeline system connecting a seller with a transmission or distribution company, we have several geographically separated markets for gas in Europe. In each market segment, trade is governed by a long-term contract, determined in negotiations between a seller and a buyer. Each contract usually stipulates the volume of gas to be delivered each year and a delivery price, which is normally linked to other energy prices. In the present chapter we confined ourselves to an atemporal analysis of some important aspects of gas contracts.

We first considered a pure bilateral monopoly, where we emphasized the nature of the buying company. The volume of trade and the range of possible negotiation outcomes for the gas price were seen to depend strongly on whether the buying company was a public utility or a profit-maximizing firm. Whereas the volume of trade was uniquely determined within each context, we were only able to derive a range for the gas price, determined as the set of prices between the agents' reservation prices.

We next considered a simple three-player bargaining game, with one buyer and two sellers. In the absence of the high-cost seller, the core of the game was identical to the price range in the pure bilateral monopoly situation. The impact of the second (high-cost) seller on the core and the (symmetric) Nash bargaining solution was that the core of the game became smaller, in the sense that the maximal price in the core now became equal to the reservation price of the second (high-cost) seller. The presence of another seller thus favours the buyer even though the buyer still purchases all gas from the low-cost seller. The appropriate Nash bargaining solution gives an equilibrium price which is equal to the Nash solution in the case with only one seller, provided this price does not exceed the upper limit of the core. If it does, the equilibrium price is equal to this upper limit, i.e. equal to the reservation price of the high-cost seller.

Finally, we presented a numerical model in which the European gas market was modelled as a game between two sellers (the USSR and Norway) and two buyers (continental Europe and the UK). The focus of attention was on production and price ranges for Norwegian gas compatible with the core of the game for the year 2010. Given the proposed assumptions about cost structure and demand, we derived a reference scenario for production in each selling country, its distribution among the buyers and a price range for Norwegian gas.

One of the most striking features of the model is the wide range of possible prices for the players which is consistent with solutions in the core. We also studied how the core depends on which assumptions we use. Norwegian exports are assumed to be determined by the Norwegian capacity limit in all of the cases we considered. However, the distribution of these exports between the UK and continental Europe depends significantly on which assumptions are used. Price and pay-off ranges are also affected by which assumptions are made. We found that the propects for Norway are worse (i) with a pipeline across the Channel than without, (ii) the lower is the demand growth, and (iii) the higher share of Soviet gas in total consumption the importing countries accept. Of these factors, the last one seems to be the most important for the price of Norwegian gas. This factor is also important for the distribution of Norwegian exports: relaxing the constraint on the share of Soviet imports gives a significant increase in the volume of gas Norway sells to the UK. We also found the somewhat surprising result that reduced costs for gas from the USSR does not unambiguously move the price range for Norwegian gas downwards: the maximal price Norway can get is higher the lower the costs of USSR gas are. In other words, the aggregate gain of reduced costs might not be divided only between the USSR and the gas buyers. Some of the aggregate gain might go to Norway, through buyers paying more for Norwegian gas although their average price for gas imports is reduced.

APPENDIX

Demand assumptions

The demand for natural gas in continental Europe and the UK is given by linear demand functions. The two parameters of each demand function in year 2010 are derived as follows: first, we find the point on the demand curve in 2010 for the 1987 real price of natural gas. This follows from the assumed GNP growth and the income elasticity of natural gas, as well as the assumed real oil price growth and the cross-elasticity of gas consumption with respect to the real oil price. In the base case GNP growth is assumed to average 2.5% throughout the period 1987–2010, both for the UK and continental Europe. The income elasticity of gas is assumed to be 0.8 in the UK and 0.7 in continental Europe. The real oil price is assumed to grow by 25% from 1987 to 2010, and the cross-elasticity is assumed to be 0.4 in both the UK and continental Europe. Once the points on the two demand curves corresponding to an unchanged gas price have been found, the two parameters of each demand function follow from the assumed direct price elasticities at these points on the demand curves. These elasticities are assumed to be -0.8 both for continental Europe and the UK.

Supply assumptions

The cost assumptions for Norway and the USSR have already been given in the text. In addition to imports from Norway and the USSR, the importing countries may have indigenous production as well as other imports. We assume that continental Europe (excluding the Netherlands) and the UK produce 25 and 30 bcm natural gas, respectively, in 2010. Moreover, it is assumed that the Netherlands and Algeria export 25 and 30 bcm to CE in 2010, respectively. The UK is assumed to import no gas from the Netherlands or Algeria, even if a pipeline across the Channel is built.

REFERENCES

Binmore, K. (1987) Bargaining models, in *Natural Gas Markets and Contracts* (eds R. Golombek, M. Hoel and J. Vislie), North-Holland, Amsterdam, pp. 239–52.

Binmore, K., Rubinstein, A. and Wolinsky, A. (1986) The Nash bargaining solution in economic modelling. *Rand Journal of Economics*, **17**, 176–88.

Hoel, M. (1986) Bargaining games with potential outside offers, University of Oslo, Department of Economics Memorandum 12/86.

Hoel, M., Holtsmark, B. and Vislie, J. (1987) The market for natural gas in Europe: the core of the game, University of Oslo, Department of Economics Memorandum 12/87.

Hoel, M. and Vislie, J. (1987) Bargaining, bilateral monopoly and exhaustible resources, in *Natural Gas Markets and Contracts* (eds R. Golombek, M. Hoel and J. Vislie), North-Holland, Amsterdam, pp. 253–65.

Hoel, M. and Strøm, S. (1987) Supply security and import diversification of natural gas, in *Natural Gas Markets and Contracts* (eds R. Golombek, M. Hoel and J. Vislie), North-Holland, Amsterdam, pp. 151–72.

Manne, A., Roland, K. and Stephen, G. (1986) Security of supply in the Western European market for natural gas. *Energy Policy*, **14**, 52–64.

Sutton, J. (1986) Non-cooperative bargaining theory: an introduction. *Review of Economic Studies*, **53**, 709–24.

Vislie, J. (1987) Long-term bilateral contracts for natural gas, in *Natural Gas Market and Contracts* (eds R. Golombek, M. Hoel and J. Vislie), North-Holland, Amsterdam, pp. 267–277.

Vislie, J. (1988) Equilibrium in a market with sequential bargaining and random outside options. *Economics Letters*, **27**, 325–8.

Vislie, J. (1989) Bargaining, vertical control, and (de)regulation in the European gas market, Chapter 4 in this volume.

4

Bargaining, vertical control, and (de)regulation in the European gas market

JON VISLIE

4.1 INTRODUCTION

In the market for natural gas in Europe, negotiations play an important role in the determination of terms of trade between various agents. Due to the small number of countries involved in the trade of nautral gas on the European continent, this market could be characterized as a 'bilateral oligopoly'. As trade between a buyer and a seller can be implemented only if the parties make investments in relationship-specific infrastructure (pipelines and terminals), the agents, once these investments have been undertaken, are locked into a bilateral monopoly position. Since the parties cannot rely on the market once the investments are undertaken, a trade agreement is usually governed by a long-term contract, established between a selling country (or a national enterprise) and a buyer, which is normally a transmission company, reselling gas to local distribution companies in various regions. Long-term contracts will not only regulate trade between upstream suppliers and transmission companies; such contracts are also prevalent in regulating trade between transmission companies and downstream firms, such as local distribution companies. Hence, there is no ordinary spot market for natural gas, but a set of segmented or geographically separated markets, where terms of trade are determined in a complex interrelated bargaining game comprising upstream as well as downstream agents. Within such a market structure, a transmission company has a rather strong market power, as a player 'in between' the upstream suppliers and the downstream buyers; for further details, see Bjerkholt, Gjelsvik and Olsen (1989).

The present paper will analyse some aspects of the European natural gas market. We focus on the aspects of bargaining or negotiations that take place between agents in a set of vertically related markets. To simplify the picture, we will consider two selling countries – Norway and the USSR – one large transmission company, which in a rough manner can be thought of as a company similar to Ruhrgas, and one local distribution company, buying natural gas from the transmission company and reselling gas to ultimate consumers, which we can think of as households. The transmission company can buy gas from either Norway, the USSR or both. The price charged by the USSR is assumed fixed, whereas the price paid for Norwegian gas is determined in negotiations. The total volume of natural gas to be purchased by the transmission company and the price will, on the other hand, be determined in negotiations with the local distribution company. With this market structure as a benchmark, we proceed to analyse aspects of vertical control and (de)regulation, in order to draw some tentative conclusions as to the impact on prices of deregulating the European gas market as proposed by the EC Commission.

According to Bjerkholt, Gjelsvik and Olsen (1989), the unit cost production and transportation of natural gas from the USSR is higher than that from Norway as of 1987. However, in the present chapter, we will assume that at least for some range of production, gas from the USSR is less expensive than gas from Norway. If the transmission company under these circumstances were allowed to buy all gas from the less expensive buyer, Norway would have been driven out of the market. However, as discussed by Hoel and Strøm (1987), the transmission company may, in order to minimize the risk of an embargo, diversify its purchases; so in that case, Norway will be able to compete, even though her gas production is more expensive. Diversification will also be the result if the recommendation given by the IEA, saying that the market share of USSR gas should not exceed 30%, is in fact obeyed. If that is the case, the transmission company would also purchase gas from Norway. In the present chapter, such recommendations are considered more closely, in order to see how the total volume of gas consumption and the prices paid between the various agents are affected by, say, an upper limit on the market share of natural gas from the USSR.

In order to cope with this rather multifaceted problem we simplify on other fronts as much as possible: we neglect that natural gas is an exhaustible resource, we analyse the whole problem within a static framework, we assume that investments in infrastructure have already been undertaken and, finally, we suppose that all bargaining outcomes satisfy the Nash bargaining solution.

4.2 THE MODEL

Let us first consider the two selling countries: the USSR is selling a volume z of natural gas to the transmission company. The (constant) unit cost of extraction and transportation is denoted c_S, and, by assumption, the price charged by the USSR per unit of z, p_S is fixed, e.g. proportional to c_S. Hence, the pay-off or net gain to the USSR from selling z units of gas to the transmission company at a net price $p_S - c_S$, is given by

$$W_S = (p_S - c_S)z \qquad (4.1)$$

The other seller, Norway, produces x units of natural gas with a (constant) unit cost of extraction and transportation c_N. The unit price paid by the transmission company is p_N, and the pay-off or net gain to Norway from selling x units at a price p_N is then

$$W_N = (p_N - c_N)x \qquad (4.2)$$

On the other side of the market, we have the local distribution company, which by assumption represents the ultimate consumers in that region or country. The gross benefit in some monetary unit of having $x + z$ units of natural gas, will be given by a strictly increasing and strictly concave benefit function $U(x + z)$. (Note that the price paid by the consumers will, by assumption, be $U'(x + z) \equiv q$. In Hoel and Vislie (1987) a model of natural gas trade for a pure bilateral monopoly situation is analysed, where the buyer is assumed to behave as a monopolist when selling gas to the ultimate consumers. This situation can be captured within the present framework by redefining the pay-off function to the local distribution company; instead of $U(x + z)$, the revenue of the monopolist would have been $U'(x + z)(x + z)$.)

In providing the ultimate consumers with natural gas, the local distribution company will usually have incurred transportation costs. However, in the present context we will neglect such costs. The price paid by the distribution company to the transmission company per unit of gas is denoted by π, which is determined in negotiations. Let us for ease of exposition assume that the local distribution company buys the entire volume of gas delivered by the two supplying countries and serves one country with natural gas. In that case, the net gain or pay-off to the consumers in this country of having $x + z$ units of natural gas at a total marginal cost equal to π, will be

$$W_B = U(x + z) - \pi(x + z) \qquad (4.3)$$

The transmission company purchases natural gas from both sellers, where deliveries from the USSR are determined from the given market share requirement, whereas deliveries from Norway follow when total demand is determined. The transfer price p_N is then determined in negotiations between Norway and the transmission company, whose gross revenue from selling $x + z$ units of gas to the buyer is $\pi(x + z)$. Hence, the transmission company's pay-off or net gain will be

$$W_T = \pi(x + z) - p_N x - p_S z \tag{4.4}$$

In order to simplify even more, we will introduce the following assumptions, for the case where contracts are determined through negotiations:

A1: The price charged by the USSR, p_S, is given.
A2: The unit cost of extraction and transportation in Norway exceeds the given price charged by the USSR, i.e. $c_N > p_S > c_S$.

Assumption A1 is introduced in order to minimize the complexity of the model. We could have allowed for negotiations between the USSR and the transmission company in the determination of p_S, but then the model would have become very complex. In order to get around the multi-player bargaining problem, we take p_S as a parameter. Assumption A2 says that the price charged by the USSR is below Norway's unit cost. This assumption can be justified on political grounds in the USSR; for instance, a high shadow price of foreign exchange might motivate the authorities in the USSR to stipulate a low value for p_S. Assumption A2 then means that if no restriction was placed upon the market share of gas from the USSR, and if there are no uncertainties of any kind, the volume of gas sold by Norway would have been zero. In order to rule out this solution, we introduce an upper limit on z as a share of $x + z$:

A3: $z \leq \alpha(x + z)$ where $\alpha \in (0, 1)$, saying that the market share for gas from the USSR should not exceed α; for instance, as proposed by the IEA, $\alpha = 0, 3$.

From the restriction in assumption A3, we can instead write $z \leq \beta x$ where the parameter $\beta \equiv \alpha/(1 - \alpha)$. As will be evident from the subsequent analysis, assumption A3 is crucial. When we go on to discuss a change in market structure, and analysing the outcome when the market segments are regulated in accordance with the proposition made by the EC Commission, we can derive explicit expressions for the loss in consumers' surplus from adhering to the policy proposed by the IEA.

In the model presented above, we have assumed away all costs of transportation. As there are likely to be decreasing unit costs of transportation due to large fixed infrastructure in the transmission sector (natural monopoly) and that the unit cost of transportation usually differs depending on where the gas comes from, we lose some interesting and important aspects of the gas market by sticking to these assumptions. However, we end up with a rather simple model where we focus mainly on price determination in the various segments of the market. A simplified picture of the structure of the European market for natural gas is presented in Fig. 4.1.

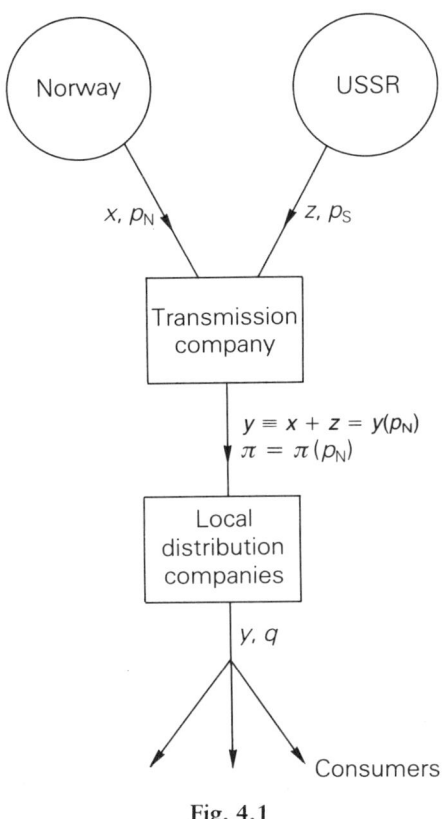

Fig. 4.1

4.3 CONTRACTS AND SUCCESSIVE BILATERAL MONOPOLIES

In this section we derive the agreements that will be established between the agents in the situation where we have successive bilateral monopolies. At each stage where negotiations take place, we assume that the resulting outcome satisfies the axioms behind the Nash bargaining solution. This means that the bargaining outcome has to be Pareto-optimal, individual rational, symmetric, invariant to linear transformations of the utility scales and independent of irrelevant alternatives; see e.g. Friedman (1986, Chapter 5).

Suppose we have the following sequence between the various negotiation stages: first, the transmission company and the local distribution company negotiate about volume of trade and price. As the outcome of these negotiations will depend upon the price paid for Norwegian gas, the local distribution company and the transmission company will agree upon a set of **contract rules**, rather than a strict contract, i.e. they will agree upon rules specifying how much gas to buy and a price π, both as functions of the price paid for Norwegian gas. Second, the transmission company and Norway negotiate about a price of Norwegian gas, given the contract rules from the first stage, and the α-requirement introduced in assumption A3 above. As the outcome of the first stage depends upon the price determined in the second stage, the strict terms of the contract between the local distribution company and the transmission company are determined once the second negotiating round is ended.

4.3.1 The bargaining outcome between the transmission company and the local distribution company

We will now consider the contract established between the transmission company and the local distribution company. They negotiate about a contract rule, expressed $\{\pi(p_N), y(p_N)\}$ where y is total gas purchase of the local distribution company, i.e. $y \equiv x + z$, and π is the price. At the point in time they negotiate, the price paid by the transmission company for natural gas delivered from Norway is not yet fixed and is considered as a parameter at this stage; so what they agree upon is a rule rather than a strict contract, where the rule is contingent upon the price paid by the transmission company for gas delivered from Norway.

According to the Nash bargaining theory, the bargaining outcome will be found as the pair $\{\pi(p_N), y(p_N)\}$ which maximizes the Nash product, i.e. the product of the parties' pay-offs, as their disagreement points, by assumption, are set equal to zero. Hence the contract will satisfy: maximize $[W_T W_B]$ with respect to y and π, under the restriction that neither of them will make a loss. In establishing these contract rules, we can solve the problem in two steps:

1. Choose a volume of trade $y(p_N)$ such that the sum of net gains is maximized.
2. Choose $\pi(p_N)$ such that total gain is distributed equally between them.

The volume of trade must satisfy the axiom of Pareto-optimality; whereas the price π according to the symmetry axiom must divide total gain equally between them. Consider the first problem:

$$\text{Maximize}_{y} \; W_B + W_T = U(y) - p_N x - p_S z$$
$$\text{s.t.} \quad z \leq \beta x \quad x \geq 0 \quad z \geq 0 \quad \text{where } y \equiv x + z \tag{4.5}$$

Suppose that $U'(0)$ is so high that the value of y that solves the problem in

Equation 4.5 is positive, even for very high values of p_N. It then follows that both y and x will be positive, and it is easy to show that $\beta x = z > 0$ as well. The optimal value of y, contingent upon p_N, must then satisfy the following condition:

$$U'(y) = (1 - \alpha)p_N + \alpha p_S \equiv q(p_N) \tag{4.6}$$

where q is the price the consumers, served by the local distribution company, are willing to pay at the margin for a total gas consumption equal to y.

From Equation 4.6 we can derive the relationship between total gas demand and consumer price, which will depend upon the price paid for Norwegian gas, p_N, as given by $y = e(q(p_N))$, of which a share $(1 - \alpha)$ is delivered from Norway. As the ultimate consumers' marginal benefit is strictly decreasing in y, we have that the lower q is, the higher is y, and due to the market share requirement on gas delivered from the USSR, the higher will x also be.

The price $\pi(p_N)$ of natural gas paid by the local distribution company to the transmission company, is determined as

$$\pi(p_N) = \arg\max_\pi [W_T W_B] \quad \text{given } z = \beta x \quad \text{and} \quad y = e(q(p_N)) \tag{4.7}$$

Due to the symmetry axiom, $\pi(p_N)$ is determined so as to make the gains to the two companies equal, i.e. $W_T = W_B$, from which it follows that the price must satisfy the following rule:

$$\pi(p_N) = \tfrac{1}{2}[\bar{U}(y) + (1 - \alpha)p_N + \alpha p_S] = \tfrac{1}{2}[\bar{U}(y) + q] \tag{4.8}$$

where $\bar{U}(y)$ is the ultimate consumers' average benefit of having y units of natural gas. (We have used Equation 4.6 in order to express π as a function of the consumer price q.)

The price π, i.e. the unit price the local distribution company is paying to the transmission company, is stipulated so as to equalize the gains to the two companies. It is obvious that $\pi(p_N)$ will exceed the consumer price q, due to strict concavity of the benefit function U. The interpretation that π exceeds the consumer price q is as follows: in order to reach an agreement with the transmission company, the consumers in the buying country as a group must collectively transfer part of their consumer surplus to the transmission company, e.g. through taxation.

It is also obvious from the expression in Equation 4.8 that π is an increasing function of p_N; the higher the price the transmission company is paying for Norwegian gas, the higher is q, and the higher q is, the smaller is y, and the higher is the average benefit to the ultimate consumers. Since $\pi(p_N)$ is differentiable, we have $\pi'(p_N) > 0$, when $\alpha \in (0, 1)$.

Above we have derived the contract rules, $\pi = \pi(p_N)$ and $y = e(q)$ with $q = q(p_N)$, not the strict terms of the contract, for the trading relationship between the local distribution company and the transmission company, as both y and π are functions of the transfer price p_N (not yet agreed upon). In order to derive strict terms of the contract established between the local distribution company and the transmission company, we have to determine the price p_N, to which we now turn.

4.3.2 The bargaining outcome for the transfer price p_N

The transmission company and the Norwegian seller will negotiate only about the price p_N. Since total demand of the ultimate consumers is determined once p_N is fixed, Norway and the transmission company can only agree upon a price, not the volume of gas trade that will take place between the two, as x is not independent of p_N. At the point in time when the transmission company and the Norwegian seller are to agree upon some price, both agents have, by assumption, perfect knowledge about the ultimate consumers' demand function; hence, for any price p_N, both agents know total demand and the relationship between the two prices, as given by $\pi = \pi(p_N)$. Inserting the demand function for Norwegian gas and the relationship between the two prices into the pay-off functions, we can express these as

$$W_N = (p_N - c_N)(1 - \alpha)e(q(p_N)) \tag{4.2'}$$

$$W_T = [\pi(p_N) - (1 - \alpha)p_N - \alpha p_S]e(q(p)) = [\pi - q]\, e(q) \tag{4.4'}$$

where we have $q = q(p_N) = (1 - \alpha)p_N + \alpha p_S$. When negotiating about the transfer price p_N, the seller knows that the higher this price is set, the higher will q and hence π be, and the lower will be the demand for natural gas. The buyer, on the other hand, is aware of the fact that the lower p_N is set, the higher will demand be, but the lower will the transfer price π also be. Both agents are therefore aware of the fact that a too favourable price, as seen from each other's perspective, will involve some opposing impact either on the volume of trade or on the price transmission company will get when reselling gas to the local distribution company. When negotiating about the price p_N, these elements are taken into account.

The price p_N is at this stage determined in such a way that the Nash product is being maximized:

$$p_N^* = \arg\max_{p_N} W_T W_N \tag{4.9}$$

In order to derive some explicit conclusions, I will from now on assume that the ultimate consumers' benefit function is given by the following constant elasticity function, with μ as the demand elasticity and A a positive constant:

$$U(y) = \frac{Ay^{(1+1/\mu)}}{1 + 1/\mu} \quad \text{where } \mu < -1 \tag{4.10}$$

Contracts and successive bilateral monopolies 75

With this benefit function, we have that the average benefit in the expression for $\pi(p_N)$ in Equation 4.8 will be equal to

$$\mu U'/(1 + \mu) = \mu q/(1 + \mu)$$

Using this in Equation 4.8, we get

$$\pi(p_N) = (1 + 2\mu)q/(2 + 2\mu)$$

When solving the problem in 4.9, which is done in the Appendix, we get the following price for Norwegian gas:

$$p_N^* = \frac{1 + 2\mu}{2 + 2\mu} c_N + \frac{\alpha}{1 - \alpha} \frac{(-1)}{2 + 2\mu} p_S \equiv \theta_1 c_N + \theta_2 p_S \quad (4.11)$$

where we have introduced the notation $\theta_1 \equiv 1 + m$, $\theta_2 \equiv \beta m$ where $m \equiv (-1)/(2 + 2\mu)$. θ_1 gives the impact on the Norwegian price of an increase in Norway's cost of production, and depends only upon the elasticity of demand; the smaller $-\mu$ is, the higher is θ_1. The coefficient θ_2 gives the relationship between the Norwegian price and the price charged by the USSR, and depends upon the market share requirement and the elasticity of demand; θ_2 is higher, the higher α (and β) is and the smaller $-\mu$ is. As $\theta_1 \equiv 1 + m$, we have $\theta_1 > 1 > \theta_2 > 0$ for finite values of μ and for $\alpha \in (0, 1)$. Norway's unit cost of production has a stronger influence on her own price than the price charged by the USSR. (We will return to Equation 4.11 in section 4.3.4.)

Inserting the expression for p_N from Equation 4.11 into Equation 4.6, we derive an explicit solution for the consumer price, q^*, as a function of the exogenous parameters

$$q^* = (1 - \alpha)p_N^* + \alpha p_S = \theta_1[(1 - \alpha)c_N + \alpha p_S] \equiv \theta_1 \Gamma \quad (4.12)$$

where Γ is defined as the term within the brackets in Equation 4.12. According to what has been said above, the price π^* follows directly as

$$\pi^* = \theta_1 q^* = \theta_1^2 \Gamma \quad (4.13)$$

With the benefit function in Equation 4.10, we have a very simple form of the demand function $y = e(q) = [q/A]^\mu$ from which we can determine the exact volume of gas purchased from the local distribution company. With the market structure characterized by successive bilateral monopolies, combined with the α-requirement on the market share of gas from the USSR, we get that total volume of gas purchase $y^* = [\theta_1 \Gamma/A]^\mu$, of which a share $(1 - \alpha)$ is delivered by the Norwegian seller.

4.3.3 The distribution of gains

We turn now to the question of how the gains from trade are distributed among the various agents, whose pay-offs are given in Equations 4.1–4.4.

On using the explicit formulae for the various prices derived in preceding sections, we find

$$W_T^* = W_B^* = mA^{-\mu}[\theta_1\Gamma]^{1+\mu} = \theta_1 W_N^* \qquad (4.14)$$

We have according to previous results that the pay-offs to the transmission and the local distribution companies are equal and it can easily be demonstrated that this (common) pay-off will exceed Norway's pay-off. The pay-off to the transmission company (and the local distribution company, which is a measure of consumer surplus) will be θ_1 times the pay-off accruing to Norway. The reason why the Norwegian seller will obtain a smaller pay-off than the buyer (the transmission company) within the present framework, is that the two agents only negotiate about price, whereas quantity follows from the ultimate consumers' demand function. As total gain to distribute between the two will now be smaller than what would have been possible is they had negotiated separately about quantity and price, the seller will 'lose' some market power to the transmission company which captures a greater share of the gain as compared to the seller; see Golombek and Vislie (1985) for further details.

4.3.4 Comparative statics

Let us consider some comparative static results for the model consisting of successive bilateral monopolies with an upper limit on the market share for gas from the USSR.

From the expression for Norway's price in Equation 4.11, we observe that p_N^* is positively related both to the market share requirement α and to the price charged by the USSR, p_S. On the other hand we note, when using assumption A2, that the consumer price q^* in Equation 4.12, and the intermediate price π^* in Equation 4.13, will be decreasing in α and increasing in p_S.

In order to understand the forces behind these relationships, let us first consider the impact of a higher maximal market share on gas delivered from the USSR. (A higher value of α may be the consequence of a change in attitude as, for example, the IEA believes that due to *perestroika*, Europe should no longer fear an embargo from the USSR.) From Equation 4.11 we note that a higher value of α, for a given p_S, yields a higher price for Norway. One explanation for this result is the following: a higher value of α will, according to Equation 4.12 and assumption A2, give a lower consumer price (Γ will be smaller) and a greater total demand. Consumption of gas delivered from the USSR will increase relatively more than total consumption, implying that total delivery from Norway will be reduced. As a result, total aggregate gain will increase. As shown in Equation 4.14, Norway's gain will be a fraction of the gain accruing to the transmission company, which is increasing in α. Since x is being reduced due to a higher value of α, Norway can obtain a higher pay-off only by being paid a higher price.

As total aggregate gain will be higher when the market share requirement is being relaxed, it seems difficult, from an efficiency point of view, to justify why the IEA should recommend its member countries to restrict their consumption of gas from the USSR. However, there are in fact other circumstances that might justify some limitation in gas purchases from the USSR, like capacity constraints and risk factors. If, for instance, α were allowed to become equal to 1, all countries, except Norway, would gain. But a modest increase in α will be to the benefit to all countries, Norway included.

Consider next the impact on the price structure of a change in the price charged in the USSR. (For instance, we can imagine the USSR becoming more 'aggressive' in the sense that p_S is being reduced, in order to sell more gas 'today' or to create some positive reputation for the future.)

Let us analyse the impact on the price structure of p_S being reduced. Less expensive gas from the USSR, with a given maximal market share α, will cause Norway's price p_N^* to fall, and the consumer price q^* will go down along with a reduction in the intermediate price π^*. As q^* is being reduced, gas consumption, consumer surplus and total aggregate gain will increase. Due to the assumption of elastic demand, Norway will sell a larger volume of gas as α is kept constant. According to how aggregate gain is divided, Norway will obtain a higher pay-off, by selling a larger volume at a somewhat lower price. Whereas the consumer price q^* and the intermediate price π^* are both unambiguously reduced as p_S becomes lower, the impact on p_N^* will in general be ambiguous, as the relationship between p_S and p_N^* is very sensitive to the formulation of the demand structure. (For instance, with a linear demand function, and inelastic demand, a lower value of p_S will cause Norway's price to go up. With inelastic demand, the only way for Norway to capture her share of a higher aggregate gain, due to a lower p_S, is by being paid a higher price.)

However, despite the ambiguity concerning the way Norway's price is being affected by a change in p_S, the more interesting consequence is that all countries gain from facing a lower price of gas delivered by the USSR.

4.4 VERTICAL CONTROL AND (DE)REGULATION

We will now consider how the proposal for deregulating the European gas market set forth by the EC Commission can be implemented. The background for such a proposal is obvious from the preceding sections, where we have seen that the transmission company is in a rather strong position *vis-à-vis* both the upstream producers and the downstream agents. This strong market power is obviously detrimental to the ultimate consumers of natural gas, in the same way as the IEA requirement on the market share of USSR gas is.

In this section we will therefore consider deregulation of the European gas market, with the purpose of reducing the power both of the transmission company and the high-cost seller, which is sheltered behind the IEA recommendation. We will analyse the proposal of the EC Commission in two steps: first, we consider the impact on prices and volume of trade by changing the vertical control in the market via backward vertical integration. Within the present context, backward vertical control means, as opposed to forward vertical integration (analysed in Dixit, 1983), that the (local) distribution company has control over the transmission company. This vertical control manifests itself through the goal of the integrated company, by serving the interests of the end users. (There are, of course, a number of problems to be solved in order to realize such a change in ownership or control. We do not have enough space to discuss all problems, but take it for granted that it is possible to change the ownership structure in the desired direction; see also Grossman and Hart (1986) for further discussion.)

Second, some deregulation might also take place among the upstream suppliers. Within the present model, deregulation might be translated into increased competition among the upstream suppliers, by abandoning the maximal market share requirement on gas from the USSR. Competition is then introduced by letting the upstream suppliers behave as duopolists in a Nash–Cournot sense when selling gas to the vertically integrated company. Instead of negotiating with the Norwegian seller, while at the same time adhering to the IEA recommendation, the vertically integrated company 'presents' its demand function for gas to the sellers, who thereafter make their quantity decisions.

Note that we introduce four separate changes in the market structure: two changes take place when the downstream companies are vertically integrated and at the same time regulated so as to behave according to the ultimate consumers' objectives. The third change takes place when we relax the market share requirement on gas delivered from the USSR, and the last one when we also allow for increased competition among the upstream suppliers. We could have analysed each change separately in order to ascribe the various effects of each change alone. However, what we will do is to consider the impact on prices, consumption and pay-offs as the two downstream companies are vertically integrated and regulated, without relaxing the market share requirement. Thereafter, we simultaneously relax the market share requirement and allow for increased competition among the upstream suppliers. Lumping both changes together can be seen as an interpretation of the proposal for deregulating the gas market. Deregulation may require the introduction of the common carrier principle, which here is interpreted literally as each supplier having 'free access' to the transmission

Vertical control and (de)regulation

network. Hence, opening up for 'free access' implies no discrimination among the sellers and increased competition in the upstream industry, and we may consider the changes simultaneously.

4.4.1 Backward vertical control with market share requirement

Let the pay-off to the vertically integrated company be denoted W_V. As this company, by assumption, is a representative for the end users, its pay-off will be equal to consumer surplus:

$$W_V = U(y) - p_N x - p_S z = U[(1 + \beta)x] - p_N x - \beta p_S x \quad (4.15)$$

where we have, according to assumption A3, that $z = \beta x$ and $y = (1 + \beta)x$. (Norway's pay-off is still given in Equation 4.2.)

The Norwegian seller and the buyer will in the new situation negotiate about a transfer price p_N and the volume of gas to be purchased from Norway. (Once the volume of gas to be purchased from Norway is determined, gas deliveries from the USSR follow directly from the market share requirement.) The two-stage bargaining structure in the previous sections is now replaced by a one-stage bargaining situation, involving only the Norwegian seller and the vertically integrated company.

In this new bargaining situation, Norway and the vertically integrated company negotiate about a transfer price p_N and a volume of trade x, so as to maximize the Nash product $W_N W_V$ with respect to p_N and x. By adopting the procedure used in section 4.3.1, the terms of the contract are now easily derived. The volume of gas to be purchased from Norway, denoted x^0 (and hence total gas consumption, denoted y^0) is determined so as to maximize the net gain to the agents. For an interior solution, we must have

$$U'(y^0)(1 + \beta) - \beta p_S - c_N = 0 \quad (4.16)$$

which can be written in a more familiar way as

$$U'(y^0) = (1 - \alpha)c_N + \alpha p_S \equiv q^0 \quad (4.17)$$

saying that the consumers' marginal benefit should be equal to a weighted average of Norway's marginal cost of production and the given price charged by the USSR, with $(1 - \alpha)$ and α as weights. This condition determines the consumer price q^0, which in the present case will be identical to the term Γ defined in Equation 4.12. When comparing the consumer price in Equation 4.17 with the one in Equation 4.12, we observe that q^0 will fall below q^*, and total gas consumption will therefore increase as a result of backward vertical control. The price paid for Norwegian gas is stipulated so as to equalize the net gains to the companies. Let this price be denoted p_N^0. Straightforward calculation yields

$$p_N^0 = \tfrac{1}{2}[\bar{U}(y^0)(1 + \beta) + c_N - \beta p_S] = \theta_1 c_N + \theta_2 p_S = p_N^* \quad (4.18)$$

where the last equality follows directly from using the benefit function in Equation 4.10.

What is surprising is that the price Norway gets is identical to the price paid to the Norweigan seller in the pre-integrated situation. As the consumer price is lower in the new situation, whereas Norway's price p_N is unchanged, Norway's pay-off, as well as consumer surplus, will increase. On using our results above, along with $y^0 = [\Gamma/A]^\mu$, the pay-offs are easily calculated:

$$W_N^0 = W_V^0 = mA^{-\mu}\Gamma^{1+\mu} > W_B^* > W_N^* \qquad (4.19)$$

The end users and the Norwegian seller will gain from the proposed change (backward vertical control) in market structure.

It is obvious that vertical integration without any changes in prices would cause no efficiency gains, along with a pay-off accruing in the new distribution company simply as the sum of the former pay-offs of the two downstream companies. However, as the vertically integrated company is negotiating directly with the Norwegian seller about a price and gas deliveries, total gain will increase as more gas will be consumed at a lower marginal price than before. (Note also that the average price paid by the consumers, which in the pre-integrated situation was identical to π, will go down as the former distortion due to the transmission company's market power vanishes.) As consumer surplus increases, Norway's net gain has to increase even more for the condition of equal division of the gains from trade in Equation 4.19 to be satisfied. This is accomplished simply by a larger volume of trade from Norway, as total consumption is higher and the market share requirement is the same as before, without changing p_N. The higher volume of trade from Norway is therefore sufficient for Norway's pay-off to increase to the same level as that of the downsteam company.

4.4.2 Backward vertical control and deregulation in the upstream industry

Consider next the following change in market structure: the vertically integrated company, which is a representative for the end users, chooses to abandon the market share requirement, by letting the suppliers in the upstream industry compete. Suppose that competition is in quantities, so the natural equilibrium concept is the Nash–Cournot equilibrium. Let again the consumer price be denoted by q. Since the vertically integrated firm is a representative for the end users, the price the consumers pay at the margin for an arbitrary volume of gas purchase y, is determined from $q = U'(y)$, from which we have the derived demand facing the duopolists, namely $q = U'(y) = P$, where P is the price paid to the sellers, determined in the Nash–Cournot equilibrium. With the benefit function in Equation 4.10, the demand function is $y = [P/A]^\mu = E(P)$, or its inverse $P = P(y) = U'(y)$.

The sellers' pay-offs or profits in the present situation are given by

$$W_N = [P(x+z) - c_N]x \quad \text{and} \quad W_S = [P(x+z) - c_S]z \quad (4.20)$$

The following two conditions constitute together a duopoly equilibrium (or a non-cooperative Nash–Cournot equilibrium) in quantities:

$$P(x+z) + xP'(x+z) = c_N \quad \text{and} \quad P(x+z) + zP'(x+z) = c_S \quad (4.21)$$

stating that each seller's perceived marginal revenue is equal to marginal cost. Adding these yields the market price P^{NC}:

$$P^{NC} = \frac{\mu}{2\mu + 1} [c_N + c_S] = \frac{\bar{c}}{1 + 1/2\mu} = q^{NC} \quad (4.22)$$

where $\bar{c} = (c_N + c_S)/2$ is the arithmetic average of the sellers' unit costs. As no transportation or transmission costs are included, the consumer price must be equal to the producer price.

In the preceding section we considered the impact of an isolated backward vertical control, without relaxing the market share requirement. We concluded that consumer price was reduced, the price the Norwegian seller obtained was unchanged, and both the Norwegian seller and the vertically integrated firm received a higher pay-off, compared with the non-integrated bargaining situation. Are these changes reinforced or counteracted when competition among the upstream suppliers is introduced?

Let us first compare the producer prices p_N^0 and P^{NC}, where p_N^0, in Equation 4.18, is the price Norway obtains in the sheltered situation. This price is influenced more by Norway's own unit cost c_N, than what the equilibrium price P^{NC} is. We have that θ_1 in Equation 4.18 will be greater than $\mu/(2\mu + 1)$, for all finite values of μ less than -1. Hence, competition among the upstream suppliers reduces the impact of the unit cost of the high-cost country, causing the producer price to go down *ceteris paribus*. On the other hand, we get a stronger influence from the USSR when competition is introduced. In Equation 4.18, the administered price p_S enters with a weight θ_2, whereas in Equation 4.22, we have c_S multiplied by $\mu/(2\mu + 1)$. It can be shown that for all values of μ less than -1.7, and for all values of α less than 0.5, this coefficient will exceed θ_2. But as $p_S > c_S$, P^{NC} might go in either direction as compared to the price Norway obtains in the 'sheltered' situation.

What can be said about the impact on the consumer price? The gas price will be reduced, *ceteris paribus*, due to the fact that the high-cost seller (Norway) loses its dominant position. On relaxing the market share requirement, via competition, high-cost gas is replaced by low-cost gas from the USSR, which *ceteris paribus*, should make it possible among the suppliers to buy a larger share of total consumption from the low-cost seller. However, the competitive effect due to duopoly, without any market power on the buyer's side, will on the other hand have a positive impact on the gas price as P^{NC} exceeds p_S. Hence, the net effect on the gas price is ambiguous.

Above we have discussed how prices are affected when opening up for some competition among the upstream producers. The next question is in what way will consumer surplus and Norway's pay-off be affected by increased competition?

Straightforward calculation yields

$$W_V^{NC} = 2mA^{-\mu}\left[\frac{\bar{c}}{1 + 1/2\mu}\right]^{1+\mu} \quad (4.23)$$

$$W_N^{NC} = \frac{2m}{1 + 2m} s_N^2 A^{-\mu}\left[\frac{\bar{c}}{1 + 1/2\mu}\right]^{1+\mu} \quad (4.24)$$

where we have used a formula, from Dixit and Stern (1982), for Norway's market share, as given by

$$s_N = -\mu + \frac{c_N[\mu + \frac{1}{2}]}{\bar{c}} \quad (4.25)$$

(An expression for the USSR's pay-off can be found in the same way as we have found Norway's pay-off; the only change is that the USSR's market share s_S, similar to s_N, is substituted for s_N in Equation 4.24.)

We found earlier that due to backward vertical control alone, consumer surplus was higher than that in the bargaining situation discussed in section 4.3, as the change in ownership or vertical control led to a lower consumer price and a lower average price. As seen from Equation 4.19, Norway will also take a share of the previous pay-off of the transmission company. Increased competition among the upstream suppliers, along with the ownership structure outlined in section 4.4.1, implied that consumer price could go either way as compared to q^0. Hence, whereas backward vertical control alone will increase both consumer surplus and Norway's pay-off, as compared to our reference solution, the impact of allowing for some competition among the upstream suppliers on consumers' net welfare is ambiguous. When comparing the consumer surplus in Equation 4.19 with that in Equation 4.23, we note that for an unchanged consumer price, consumer surplus will double due to competition. However, as noted above, competition alone may cause the consumer price to go in either direction. Hence, there will be an offsetting effect on consumers' welfare if the consumer price should increase and an additional positive effect if it should decrease.

It is also obvious that the USSR will experience an increase in the pay-off W_S, as both price and quantity sold from the USSR increase. As mentioned above, Norway's pay-off increased as downstream firms integrated vertically. From Equation 4.24 we observe that even with the consumer price unchanged, competition will have a negative impact on Norway's pay-off. If competition should also lead to a higher consumer price, Norway would suffer an even greater loss.

Appendix

4.5 CONCLUSIONS

We have considered the market for natural gas in Europe in a number of highly stylized models. We started out with a model emphasizing the role of bargaining in a set of vertically related markets, when there was assumed to be a restriction on the market share of gas delivered from the USSR, which was assumed to be the low-cost country. The various prices were derived under the assumption of a Nash bargaining solution. In the first situation, where Norway was sheltered by a market share requirement on gas from the USSR, we found that the transmission company had a strong market power. Negotiation between the local distribution company and the transmission company on the one hand, and negotiation between the Norwegian seller and the transmission company on the other, gave the transmission company a rather strong position both upstream and downstream. The bargaining outcome led to the equal pay-off to the downstream companies, but in excess of the pay-off to Norway. We then considered the impact of changing the vertical ownership structure (backward vertical control), while maintaining the market share requirement on gas from the USSR. The result of changing the vertical ownership structure was that the consumer price was reduced, whereas both Norway and the consumers gained from this change. Finally, we considered the impact of introducing some competition among the upstream suppliers. We derived the market solution (Nash–Cournot equilibrium) and found that the impact on the price structure was ambiguous. Even though the impact on consumers' surplus was ambiguous, it was obvious that the USSR will have a substantial gain from more competition, whereas Norway will lose.

APPENDIX

We will in this Appendix derive the price Norway obtains in the constrained bargaining situation, i.e. p_N^*, together with the consumer price q^* and the intermediate price π^*.

The price p_N^* was the solution of the following problem:

$$p_N^* = \arg\max_{p_N} [W_N W_T] \tag{A.1}$$

where the pay-offs to Norway and the transmission company, W_N and W_T, are given by

$$W_N = (p_N - c_N)x = (p_N - c_N)(1 - \alpha)e(q(p_N)) \tag{A.2}$$

$$\begin{aligned} W_T &= \pi y - p_N x - p_s z = \pi(1 + \beta)x - p_N x - \beta x p_s \\ &= [\pi(p_N)(1 + \beta) - p_N - \beta p_s](1 - \alpha)e(q(p_N)) \end{aligned} \tag{A.3}$$

where we have used that $z = \beta x$ and $x = (1 - \alpha)e(q)$, where $e(q)$ is the ultimate consumers' demand function for natural gas.

Let $N(p_N) \equiv W_N W_T$. The transfer price p_N^* is then determined from setting $N'(p_N) = 0$. Differentiation with respect to p_N yields, when we use that $e'(q)$ and $q'(p_N)$ are the derivatives of $e(q)$ and $q(p_N)$ with respect to q and p_N respectively:

$$N'(p_N) = W_N[x(\pi'(p_N)(1 + \beta) - 1) + W_T(1 - \alpha)e'(q)q'/x]$$
$$+ W_T[x + W_N(1 - \alpha)e'(q)q'/x] = 0 \quad (A.4)$$

Now using that $q' = 1 - \alpha$ from Equation 4.6, the expression for $N'(p_N)$ can be written as follows:

$$[W_T - W_N]x + W_N x \pi'(1 + \beta) + 2W_T W_N(1 - \alpha)^2 e'/x = 0 \quad (A.5)$$

From the text we have the demand elasticity $\mu \equiv E\, ly : q = qe'/y$, which is being used in order to eliminate $e'(q)$. In addition we have $\pi^* = \pi(p_N)$, defined in Equation 4.8. We can then use Equation 4.8 to find an explicit expression for $\pi'(p_N)$. Straightforward differentiation yields

$$\pi'(p_N) = (1 - \alpha)[1 + \mu(1 - \bar{U}/U')]/2 \quad (A.6)$$

Using that $U'(y) = q$ and the fact that with the benefit function in Equation 4.10 we have $\bar{U} = [\mu/(1 + \mu)]U'$, we can write Equation A.6 as

$$\pi'(p_N) = (1 - \alpha)(2\mu + 1)/(2\mu + 2) = (1 - \alpha)\theta_1 \quad (A.6)'$$

where we have used the definition of θ_1 in Equation 4.11. With the constant elasticity benefit function in Equation 4.10, we can find an explicit solution for the transfer price π^*:

$$\pi^* = \pi(p_N) = \theta_1 q(p_N) \quad (A.7)$$

as given in Equation 4.13. Inserting for $\pi'(p_N)$ into Equation A.5, and using that $(1 + \beta)(1 - \alpha) = 1$, and dividing through by W_N/x yields

$$[W_T - W_N]/W_N + \theta_1 + (1 - \alpha)^2 2W_T e'/x^2$$
$$= W_T/(W_N) - 1 + \theta_1 + 2(1 - \alpha)\mu W_T/qx = 0$$
$$(A.8)$$

Using the definition W_T and W_N in Equation A.8, we find

$$(\pi - q)/[(p_N - c_N)(1 - \alpha)] + \theta_1 - 1 + 2(\pi - q)\mu/q = 0 \quad (A.9)$$

Inserting from Equation A.7 and some algebra yields

$$q + (1 + 2\mu)(1 - \alpha)(p_N - c_N) = 0 \quad (A.10)$$

Using the definition of q from Equation 4.6, we easily derive the desired result:

$$p_N^* = \theta_1 c_N + \theta_2 p_S \qquad (A.11)$$

and the consumer price follows directly when inserting Equation A.11 into the definition 4.6. We then get

$$\begin{aligned} q^* &= (1 - \alpha)[\theta_1 c_N + \theta_2 p_S] + \alpha p_S = (1 - \alpha)\theta_1 c_N + \alpha \theta_1 p_S \\ &= \theta_1[(1 - \alpha)c_N + \alpha p_S] \equiv \theta_1 \Gamma \end{aligned} \qquad (A.12)$$

ACKNOWLEDGEMENTS

The author gratefully acknowledges the constructive comments of Olav Bjerkholt, Michael Hoel, Einar Hope, Øystein Olsen and Robert Weiner on an earlier draft.

REFERENCES

Bjerkholt, O., Gjelsvik, E. and Olsen, Ø. (1989) Gas trade and demand in Northwest Europe: regulation, bargaining and competition, Discussion Paper; no. 45, Central Bureau of Statistics, Oslo, Norway.
Dixit, A. (1983) Vertical integration in a monopolistically competitive industry. *International Journal of Industrial Organization*, 1, 63–78.
Dixit, A. and Stern, N. (1982) Oligopoly and welfare. A unified presentation with applications to trade and development. *European Economic Review*, 19, 123–43.
Friedman, J. W. (1986) *Game Theory with Applications to Economics*, Oxford University Press.
Golombek, R. and Vislie, J. (1985) On bilateral monopoloy – a Nash–Wicksell approach, Memorandum from Department of Economics, University of Oslo, 23 January.
Grossman, S. J. and Hart, O. D. (1986) The costs and benefits of ownership: a theory of vertical and lateral integration. *Journal of Political Economy*, 94, 691–719.
Hoel, M. and Strøm, S. (1987) Supply security and import diversification of natural gas, in *Natural Gas Markets and Contracts* (eds R. Golombek, M. Hoel and J. Vislie), Elsevier Science Publishers B. V. North-Holland, Amsterdam, pp. 151–72.
Hoel, M. and Vislie, J. (1987) Bargaining, bilateral monopoly and exhaustible resources, in *Natural Gas Markets and Contracts* (eds R. Golombek, M. Hoel and J. Vislie), Elsevier Science Publishers B. V. North-Holland, Amsterdam, pp. 253–65.

5

Environmental effects of a transition from oil and coal to natural gas in Europe

KNUT H. ALFSEN, LORENTS LORENTSEN AND

KARINE NYBORG

5.1 INTRODUCTION

The last few years has brought a growing recognition of and concern about the large-scale environmental problems caused by the massive use of fossil fuels. International problems with acid rain have been brought up to high political levels on a number of occasions (Norway vs the UK, Canada vs the USA). Popular articles, conferences, commissions and committees addressing the greenhouse problem have flourished. The discovery of the Antarctic ozone hole, the probable verification of global warming, damage such as dead forests in the wake of acid rain, the continuous effort of serious environmental pressure groups and the adoption of environmental problems high on the agenda of the UN (cf. The World Commission on the Environment and Development, 1987) have all played an important part in enhancing awareness of threatening environmental problems.

The aim of this chapter is to examine some environmental consequences of a European switch from oil and coal to natural gas as a fuel, particularly the effects of SO_2 and CO_2 emissions and sulphur deposition. We are not making any attempt to investigate fully the best ways to reduce or 'eliminate' pollution of these kinds. The environmental effects of the use of natural gas as a fuel are thus not compared to the effects of energy sources other than fossil fuels, such as hydro or nuclear power.

The possible reductions in SO_2 and CO_2 emissions by a transition to natural gas are estimated in this chapter using available data on 1985 emissions and energy consumption. By means of a transportation matrix, the reductions in SO_2 emissions are then translated into reductions of sulphur deposition in various European regions. The reduction in emissions (and sulphur deposition) is calculated under alternative assumptions as to what degree various sectors of the economy can change from oil and coal to natural gas, keeping the total utilized energy consumption constant. Thus, we ignore the possible effect of lower total energy consumption due to the substitution of low-priced fuels with a higher-priced energy commodity. For the substitution actually to occur, the switch must be technically feasible and economically viable. A comprehensive evaluation of technical and economic viability by industry and country would require more detailed data than available for this study. There are also other data problems described in more detail in the following, and the estimates presented below must be considered preliminary.

Acid rain and the greenhouse effect should be a concern for all nations. Norway, in particular, with its huge reserves of natural gas, should consider the potential environmental benefits of a large-scale transition from coal and oil to natural gas where possible. These benefits, associated with the low sulphur content and high energy content of natural gas compared to other fossil fuels, should be taken into account in the long-term planning of the exploitation of natural gas and in the negotiation of contracts for deliveries of natural gas.

Large-scale transition from oil and coal to gas in stationary combustion of fossil fuels will reduce SO_2 emissions in Europe by approximately 90%. A switching of fuels in power plants alone will reduce the emissions by more than 50%. Reductions in deposition of sulphur will vary among countries, but generally be of the order of 30–60% in the case when fuel switching takes place in power plants only, and from roughly 70 to almost 100% in the case where all stationary combustion of fossil fuels is based on gas (Fig. 5.1).

The effect on CO_2 emissions from a total transition to natural gas is far less dramatic than for SO_2. Still, a potential reduction of approximately 30% in European OECD countries is indicated by the calculation when all oil and coal in stationary combustion is substituted by gas. The absolute reduction is largest in countries where Norway already has relatively large shares of the gas market (West Germany and the UK).

Expansion of economic activity will—*ceteris paribus*—lead to a growing demand for fossil fuels and to an increase in CO_2 emissions over time. According to our calculations, half of the projected consumption of oil and coal in European OECD countries in 1995 must be substituted by gas just to keep CO_2 emissions down to the 1985 level.

Concerning the supply of natural gas, it seems possible that production in north-west Europe alone can cover 50% of the demand to substitute oil and coal outside the transportation sector (Odell, 1988).

Effects on SO_2 emission and deposition of a transition to gas

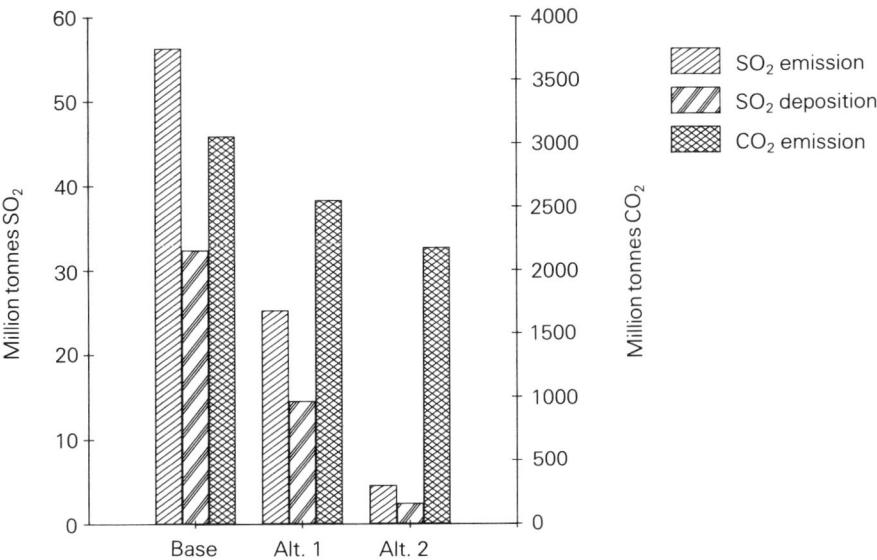

Fig. 5.1 Effects of a transition to natural gas in Europe.

5.2 EFFECTS ON SO_2 EMISSION AND DEPOSITION OF A TRANSITION FROM OIL AND COAL TO GAS

5.2.1 SO_2 data

Data on the 1985 emissions of SO_2 in European countries are taken from the RAINS project (Alcamo et al., 1986). These are projections based on official statistics of SO_2 emissions in 1980 and forecasted growth in energy consumption, technological changes and investment in cleaning equipment. Uncertainties in official statistics and in the forecasts from 1980 to 1985 will of course be reflected in the 1985 data.

Anthropogenic (man-made) emissions of SO_2 can be divided into two groups:

1. Emission from combustion of oil and coal. A transition from these fuels to natural gas will virtually eliminate the SO_2 emissions. In this study only emissions from stationary sources will be considered.
2. Emissions related to other industrial processes than combustion of sulphur-rich fuels (e.g. emission from the cathodes in the electrolysis of aluminium). These process emissions will not be affected by a transition to natural gas in stationary combustion.

In most European countries combustion-related emissions are the major source of SO_2. However, in Norway approximately half of the SO_2

emissions are process related. This is due to the fact that almost 100% of the electricity consumption in Norway is hydro power, of which one-third is used to produce aluminium, ferro alloys and pulp and paper.

The reference case of SO_2 emissions are the 1985 emissions as reported by the Regional Acidification Information and Simulation Model (RAINS) project. According to the RAINS data, process emissions are zero in all East European countries as well as in Denmark, Greece, Ireland, Luxemburg, Portugal, Switzerland, Turkey and Yugoslavia. This is obviously not correct, and bias the calculated percentage reduction in SO_2 emissions.

Two alternatives for substituting natural gas for oil and coal are considered. Alternative 1 is obtained by assuming total replacement of coal and oil by natural gas in all power plants in Europe. Alternative 2 is obtained by assuming a total transition from oil and coal to natural gas in all sectors, except for the transportation sector. Note that the individual country results in alternative 2 are strongly influenced by missing data on process emissions. Only the results for Europe in total are therefore reported in the tables.

The two alternatives are chosen somewhat arbitrarily. They can nevertheless be useful in providing estimates of maximum limits of achievable gains from a switch to gas. Note, however, that some transition to gas would be technically possible in the transportation sector as well. Also note that the relationship between SO_2 emissions and the degree of substitution is not necessarily linear. Substituting coal and oil in the plants with no cleaning equipment would yield great reductions per tonne of fuel than transitions in plants with efficient cleaning.

Sulphur deposition is calculated by multiplying the emissions with a transportation matrix. The transportation matrix is based on a detailed model of sulphur transport in the atmosphere developed by the Western Meteorological Synthesizing Centre (MSC-W) of EMEP (Programme for Monitoring and Evaluation of the Long Range Transmission of Air Pollutants in Europe)(Eliassen and Saltbones, 1983; Eliassen, Lemhaus and Saltbones, 1986). The transporation matrix employed in this study is an average of the model calculation for each of the years 1979, 1980, 1983 and 1984 (Nyborg, 1987). The matrix thus represents some sort of average expected transportation pattern. No account is taken for stochastic or uncertain elements in the transportation process.

Part of the deposits of sulphur comes from sources that are unidentifiable within the framework of the tranportation model, either because the sources are located outside of Europe or because the residence time of the sulphur in the atmosphere is too long to be captured by the model calculations. This part of the total deposition is larger at the fringe of Europe (e.g. Norway where roughly half of the deposition is from unknown sources) than in more central regions. It is reasonable to believe that at least a part of the unidentifiable depositions will be reduced by a transition to natural gas.

Effects on SO_2 emission and deposition of a transition to gas

However, this has not been modelled in the present study. Only the reduction in deposition due to reduction in emissions from known sources are considered. This leads to an underestimation of the absolute reductions in total sulphur depositions in all countries, and will be more important in remote countries like Norway.

5.2.2 Results

In the alternative where oil and coal used for power generation is replaced with natural gas (alternative 1), there is a considerable reduction in emissions and deposition; 55 and 48%, respectively, for total Europe (reduction in deposition is measured relative to deposition including contributions from unknown sources). Reduction in Norwegian emissions are almost zero (due to the high percentage of hydro power in Norway) in this case, but depositions are reduced by almost 25%. Deposition in Norway from identifiable sources is reduced by approximately 50%.

As expected, the reduction of SO_2 emissions and depositions are even greater in the scenario where all use of oil and coal in stationary sources is substituted with non-sulphurous natural gas (alternative 2). The total reduction in emissions is calculated to be 92% for total Europe relative to the base case scenario. Depositions are reduced by 80% when the fixed deposition level due to unknown sources is included. The deposition in Norway originating from known foreign sources will almost disappear, but Norwegian process emissions will continue to contribute to the national deposition. Tables 5.1 and 5.2 show the details of the effects on SO_2 emissions and depositions for alternative 1 and total change for alternative 2.

For Norway, Sweden and Finland, the reduction of SO_2 emissions in other countries is clearly essential if depositions are to be reduced significantly. This is demonstrated by the calculations above, which in fact underestimate the importance of reductions in emissions in other countries, since the deposits from unidentified sources are kept constant. Thus, it is clearly in the interest of the Nordic countries to promote the use of natural gas in the rest of Europe, in order to limit the damage caused by deposition of sulphur.

5.2.3 Future outlook

The major part of SO_2 emissions can be eliminated by installing cleaning equipment. Hence, future developments in sulphur emissions do not depend upon energy consumption and activity levels in industries with process-related emissions alone, but also on the extent to which appropriate cleaning measures are taken in each country.

Table 5.1 Emission of SO_2 (1000 tonnes) and percentage change relative to base scenario

	Base	Alternative 1	Alternative 1	2
	(1000 tonnes)		(% change)	
Albania	112	49	−56	
Austria	295	225	−24	
Belgium	587	408	−30	
Bulgaria	1 026	522	−49	
Czechoslovakia	3 080	1 196	−61	
Denmark	394	187	−53	
Finland	466	422	−9	
France	2 301	1 759	−24	
Germany, East	4 719	1 442	−69	
Germany, West	2 951	1 154	−61	
Greece	828	226	−73	
Hungary	1 528	554	−64	
Ireland	148	115	−22	
Italy	2 988	1 771	−41	
Luxemburg	36	33	−8	
Netherlands	293	188	−36	
Norway	125	124	−1	
Poland	4 211	1 218	−71	
Portugal	309	192	−38	
Romania	1 442	641	−56	
Spain	3 130	1 880	−40	
Sweden	434	398	−8	
Switzerland	100	97	−3	
Turkey	1 120	791	−29	
UK	4 229	1 602	−62	
USSR	18 018	7 501	−58	
Yugoslavia	1 478	630	−57	
Total	56 348	25 325	−55	−92

Source: Central Bureau of Statistics of Norway, Oslo, RAINS.

Most European countries have signed an agreement to reduce SO_2 emissions by at least 30% from 1980 to 1993 (UN/ECE, 1985). However, as economic growth is expected to continue, it is reasonable to believe that the activity level in SO_2 emitter sectors will increase. Over time, more efficient cleaning measures will then be required to keep emission levels down. Model calculations from the RAINS project indicate that SO_2 emissions from European countries were reduced by approximately 5% from 1980 to 1985, but are predicted to rise by 4% from 1985 to 1990.

If emissions were to develop at the same speed as GDP, consequences would be more dramatic. Assuming a development in SO_2 emissions from

Effects on SO₂ emission and deposition of a transition to gas

Table 5.2 Deposition of sulphur (1000 tonnes) and percentage change in total deposition relative to base scenario

	Base scenario: deposition sources[a]		Alternative	
	Total (1000 tonnes)	Unknown	1 (% change)	2 (% change)
Albania	43	11	−41	
Austria	231	28	−44	
Belgium	128	9	−36	
Bulgaria	265	26	−47	
Czechoslovakia	775	29	−60	
Denmark	104	12	−50	
Finland	286	80	−28	
France	902	164	−29	
Germany, East	890	24	−65	
Germany, West	949	74	−52	
Greece	202	30	−55	
Hungary	377	20	−58	
Ireland	55	23	−22	
Italy	726	90	−40	
Luxemburg	8	1	−31	
Netherlands	144	12	−45	
Norway	218	107	−24	
Poland	1378	64	−64	
Portugal	88	26	−27	
Romania	571	55	−52	
Spain	665	102	−34	
Sweden	360	111	−30	
Switzerland	96	16	−31	
Turkey	347	84	−29	
UK	718	66	−55	
USSR	4794	755	−49	
Yugoslavia	604	81	−48	
Total	15918	2100	−48	−80

[a] Arithmetic mean of deposition in 1979, 1980, 1983 and 1984. Kept constant in all alternatives.
Source: MSC-W, Central Bureau of Statistics of Norway, RAINS.

1980 to 1985 as projected by RAINS, and thereafter growth rates of SO₂ emissions equal to the growth rates of GDP, yields a 12% increase in emissions from 1980 to 1990. (Economic growth rates are based on the international economic modelling project LINK, as forecasted by the LINK spring meeting 1987 (see also Klein, 1977). For East European countries, growth rates of net material product (NMP) were employed.)

To achieve the goal of a least 30% reduction in emissions from 1980 to 1993, serious efforts to reduce the ratio between emissions and GDP (NMP)

94 *Effects of a transition from oil and coal to natural gas*

will have to be made. A switch from oil and coal as fuel to natural gas is one possible way to do this. When considering which cleaning measures to choose, the costs of cleaning ought to be compared to the costs of switching to gas where this is possible.

5.3 EFFECTS ON CO_2 EMISSIONS BY A TRANSITION FROM OIL AND COAL TO GAS

5.3.1 CO_2 Data and method

A switch from combustion of oil and coal to natural gas, keeping the level of utilized energy fixed, will reduce the emissions of CO_2 as well as SO_2. The calculations which follow were based on detailed energy statistics by fuel type, sector and country, information which was not available for East European countries. The CO_2 calculations are therefore confined to the European OECD countries (also including Yugoslavia), and are based on OECD/IEA's energy balances for 1985 (OECD/IEA, 1988a).

Only CO_2 emissions from the combustion of oil, coal and gas are included. Data on consumption of other solid fuels are given in the energy balances, but are not included in the estimates of CO_2 emissions. CO_2 emissions from ocean transport and emissions from other anthropogenic sources than combustion are also excluded.

The CO_2 combustion emissions are calculated for the following sectors:

1. power plants
2. industry
3. transport
4. other sectors

Emissions are calculated according to the following expression:

$$E_{ij} = X_{ij} * V_{ij} * U_{ij}$$

where E_{ij} is the emission of CO_2 in 1000 tonnes from combustion of fuel i in sector j, X_{ij} the consumption of fuel i in sector j measured in theoretical energy content of the fuel, V_{ij} the fraction of useful energy to theoretical energy content of fuel i in sector j (energy efficiency coefficient) and U_{ij} the emission of CO_2 per unit of useful energy from combustion of fuel i in sector j. The coefficients V_{ij} and U_{ij} are assumed equal in all countries concerned and are based on Rosland (1987). Table 5.3 shows the actual values used.

5.3.2 Results

Tables 5.4 and 5.5 show the effect of substituting a fraction f of the oil and coal consumption in the different sectors by natural gas, keeping total useful

Effects on CO_2 emissions by a transition from oil and coal to gas 95

Table 5.3 Fuel efficiency and emission per unit utilized energy

	Fuel efficiency (V_{ij})			Emission coefficient (U_{ij})		
	Coal	Oil	Gas	Coal	Oil	Gas
					(tonnes CO_2 TJ^{-1})	
Power plants	0.40	0.40	0.47	270	190	120
Industry	0.80	0.85	0.80	135	88	70
Transport	—	0.25	—	—	302	—
Other sectors	0.60	0.70	0.75	179	108	75

Source: Central Bureau of Statistics of Norway.

Table 5.4 Emission of CO_2 due to combustion of oil, coal and gas in European OECD countries [a] in 1985 (million tonnes CO_2)

	Substitution fraction (f)			
	0	0.1	0.5	1.0
Power plants	950	903	715	481
Industry	666	647	569	473
Transport[b]	695	695	695	695
Other sectors	750	731	655	560
Total	3061	2976	2635	2208

[a] Also including Yugoslavia.
[b] No subsitution.

Table 5.5 Percentage reduction in CO_2 emissions from European OECD countries [a]

	Substitution fraction (f)		
	0.1	0.5	1.0
Power plants	5	25	49
Industry	3	15	29
Transport[b]	0	0	0
Other sectors	3	13	25
Total	3	14	28

[a] Also including Yugoslavia.
[b] No subsitution.

energy consumption fixed. In transportation no substitution is assumed. The transportation sector includes all transport except international marine bunkers.

Effects of a transition from oil and coal to natural gas

The largest reduction is within the power-generating sector, in particular in countries which rely on coal-fired power plants.

Among the European OECD countries, the UK and West Germany will both experience especially large reductions in CO_2 emissions when switching from oil and coal to gas. Both countries have power production based largely on coal-fired power plants. From a Norwegian point of view it is interesting that these countries, which are the two major buyers of Norwegian gas, also represent the countries where the largest absolute environmental benefits (including lower deposits of sulphur) from fuel switching can be realized.

France and Italy can also achieve large reductions in CO_2 emissions by switching fuel. Both of these countries use coal extensively in their industrial production.

There are large variations in CO_2 emission per capita among the countries listed in Table 5.6. The countries with annual CO_2 emissions from

Table 5.6 Emission of CO_2 from combustion of fossil fuels,[a] 1985. Reductions when oil and coal are replaced by gas in all sectors except transport

	Emission (million tonnes)	Reduction (million tonnes)	Reduction (%)	Emission per capita	Reduction per capita (tonnes)
Austria	53	13	−25	7.0	1.7
Belgium	100	28	−28	10.1	2.8
Denmark	66	24	−36	12.9	4.7
Finland	45	13	−30	9.1	2.7
France	374	83	−22	6.8	1.5
Germany, West	734	237	−32	12.0	3.9
Greece	60	20	−33	6.0	2.0
Iceland	2	0.2	−12	8.0	1.0
Ireland	23	5	−23	6.5	1.5
Italy	343	73	−21	6.0	1.3
Luxemburg	10	4	−34	28.2	9.5
Netherlands	146	19	−13	10.1	1.3
Norway	25	5	−19	5.9	1.1
Portugal	25	5	−22	2.5	0.5
Spain	182	55	−30	4.7	1.4
Sweden	60	14	−23	7.2	1.7
Switzerland	43	7	−17	6.8	1.2
Turkey	96	33	−35	1.9	0.7
UK	562	169	−30	9.9	3.0
Yugoslavia	112	45	−40	4.9	1.9
OECD Europe	3061	853	−28	7.2	2.0

[a] Excluding ocean transport.

Effects on CO_2 emissions by a transition from oil and coal to gas

combustion of fossil fuels of approximately 10 tonnes per capita or more are Luxemburg, Denmark, West Germany, the Netherlands, Belgium and the UK. Most of these countries have coal-fired power plants and production of iron and steel based on coal as fuel. The exception is the Netherlands which already base much of their power and industrial production on natural gas. Approximately 50% of the calculated CO_2 emission from combustion of fossil fuels was due to burning of gas in the Netherlands in 1985. This shows that even though gas is a better fuel that oil and coal with respect to the greenhouse problem, combustion of gas still releases large amounts of CO_2.

5.3.3 Forecasts to 1995

Unlike SO_2, there are as yet no practically and economically feasible techniques to remove or 'clean' vast amounts of CO_2 gas. Emission levels are thus closely connected with combustion of fossil fuels.

We have estimated CO_2 emission levels in 1995 based on energy consumption forecasts presented by the OECD/IEA (1988b). These energy forecasts reflect government energy projections in each of the IEA member countries. For the IEA non-member countries Finland, France and Yugoslavia, the calculations are based upon ECE data (UN/ECE, National and Regional Energy Balances). The ECE forecasts were only available for the years 1990 and 2000. To obtain estimates of energy consumption in 1995 arithmetic means of the numbers predicted for 1990 and 2000 were employed. Energy forecast for Iceland was not available; hence numbers for Iceland are kept constant at the 1985 level.

Fuel efficiency is assumed constant from 1985 to 1995. Technical progress which makes fossil fuel combustion more effective is thus not considered. Growth rates of non-energy use of fossil fuels are assumed equal to the growth rates of total energy demand of each sector respectively.

In the base case scenario for 1995, both fuel mix and total fuel consumption are equal to the IEA and ECE projections. The calculations show that CO_2 emissions from the European OECD countries (also including Yugoslavia) can be expected to increase by 14% from 1985 to 1995. About 85% of this increase is due to growth in emissions from combustion of coal. Emissions from power plants increase by 30%, from industry and transport by 12–15%, while emissions from other sectors decrease by 6% (Table 5.7).

Relative increases are largest for Turkey, Portugal and Yugoslavia (55–130%), which were among the countries with lowest emissions per capita in 1985. Each of these countries project to increase their consumption of fossil fuels by more than 50%, according to the IEA and ECE date. Turkey, the UK, Yugoslavia and Italy will experience the largest absolute growth in emission levels (60–125 million tonnes each). The countries mentioned

Table 5.7 Percentage change in CO_2 emissions from European OECD countries[a] 1985–1995

	Substitution fraction (f)			
	0	0.1	0.5	1.0
Power plants	+30	+24	− 3	−36
Industry	+12	+ 9	− 5	−21
Transport[b]	+15	+15	+15	+15
Other sectors	− 6	− 9	−17	−28
Total	+14	+11	− 3	−19

[a]Also including Yugoslavia.
[b]No substitution.

above all have in common the fact that they project considerable increases in their production of electricity in coal-fired power plants.

The only country with a significant reduction in the estimated CO_2 emission level is France (24%). According to the ECE data, France will decrease its use of coal for power production by more than 50% and rely more heavily on nuclear power.

The emission levels are projected to increase partly because of a switch from oil to coal in many countries. But the main cause is that total demand for fossil fuels is growing. Total consumption of fossil fuels, measured in tonnes of oil equivalent, is expected to grow by 12% from 1985 to 1995.

A further transition to natural gas from coal and oil would, however, reduce the effects of increasing energy demand on emission levels. Table 5.7 shows relative changes in CO_2 emission levels from 1985 to 1995 under different assumptions of transition to gas. A substitution fraction f means that a fraction f of the projected consumption of oil and coal in 1995 is replaced by natural gas, while the forecasted level of delivered energy in 1995 is kept constant. In transportation no substitution is assumed.

Because of the increase in consumption of fossil fuels from 1985 to 1995, a substitution fraction of about 0.5 is required to keep emissions at the 1985 level. In other words, half of the projected use of oil and coal in sectors other than transportation must be replaced by natural gas to avoid increased CO_2 emissions.

The distribution of gas to local consumers requires vast investment (pipelines, etc.). The extent to which this infrastructure already exists varies between countries. However, the production of electric power is usually undertaken in large utilities, and for this reason the distribution costs might be a less severe problem in this sector than for the industry and household sectors. In fact, if all coal- and oil-fired power plants switched to natural gas, this would imply a 7% reduction of the total CO_2 emissions from 1985 to 1995 instead of the forecasted 14% increase. This reduction would be despite the increased energy consumption and would require no further transition to gas in other sectors compared to the IEA/ECE forecasts. Note that the effects on CO_2 reduction would be much greater if gas were used directly by end users and not converted to electricity (at an average efficiency rate of 0.47).

5.3.4 How much gas will be needed?

The calculations above would only be meaningful if there is gas supply available to substitute oil and coal for a relatively long period. In the 1995 IEA forecasts, coal and oil consumption in all sectors except transportation in European IEA countries are 340 million tonnes oil equivalent (mtoe) coal and 208 mtoe oil. Roughly estimated, a transition to gas would require 500 bcm natural gas. This is approximately twice the current consumption of gas in Western Europe.

5.4 THE IMPACT ON GAS CONTRACT NEGOTIATIONS

Reductions in SO_2 and CO_2 emissions are in the interest of all European nations, including Norway as a major receiver of acid deposition. This may affect the Norwegian, as well as the Soviet and Algerian, attitude in future negotiations of gas contracts.

Combustion of natural gas has negative external effects, releasing gases which contribute to the acid precipitation problem in areas far away from the combustion site (NO_x) and contribute to the greenhouse effect (CO_2). The negative external effects are, however, less severe than those of the consumption of coal and oil. Hence, positive environmental effects of a transition to natural gas as a fuel are inevitably linked to reductions in the combustion of coal and oil. Attempting to accelerate this transition by lowering the gas price might eventually lead to growth in the amount of unwanted emissions, due to increases in the total consumption of fossil fuels.

The question of how to deal with environmental issues in gas contract negotiations needs further consideration. Data on emissions and deposition of energy-related pollutants are clearly relevant as background material in gas contract negotiations, and efforts should be made to bring forward such information.

REFERENCES

Alcamo, J., Hordijk, L., Kämäri, J., Kauppi, P., Posch, M., and Runca, E., (1986) *Integrated Analysis of Acidification in Europe*, Research Report 2/86, IIASA, Luxemburg.

Eliassen, A. and Saltbones, J. (1983) Modelling of long-range transport of sulphur over Europe: a two-year model run and some model experiments. *Atmospheric Environment*, 17.

Eliassen, A., Lemhaus, J. and Saltbones, J. (1986) *A modified sulphur budget for Europe for 1980*, EMEP/MSC-W Report 1/86, The Norwegian Institute of Meteorology, Oslo, pp. 1457–73.

Klein, L.R. (1977) *Project LINK*, Lecture Series 30, Centre of Planning and Economic Research, Athens.

Nyborg, K. (1987) Økonomisk aktivitet og utslipp av svovel i Europa. Utslippsframskrivninger med basis i LINK-prosjektets økonomiske framskrivninger (Economic activity and emission of sulphur in Europe. Forecasts based on economic growth paths from the LINK-project), Interne Notater 87/37, Central Bureau of Statistics of Norway, Oslo.

Odell, P. R. (1988) The West European gas market. The current position and alternative project *Energy Policy*, **16**, 480–93.

OECD/IEA (1988a) *Energy Balances of OECD Countries 1985/86*, Paris.

OECD/IEA (1988b) *Energy Policies and Programmes of IEA Countries. 1987 Review*, Paris.

Rosland, A. (1987) Utslipp av karbondioksid fra antropogene kilder (Emission of carbon dioxide from anthropogenic sources), unpublished note, Central Bureau of Statistics of Norway, Oslo.

RAINS model version 3.

UN/ECE (1985) *Protocol to the 1979 Convention on long-range transboundary air pollution on the reduction of Sulphur Emissions or their Transboundary Fluxes by at least 30 per cent*. ECE/EB.AIR/12.

UN/ECE (1988) *National and Regional Energy Balances for Europe and North America*, Geneva.

The World Commission on Environment and Development (1987) *Our Common Future*, Oxford University Press, Oxford, New York.

Part Two
Management of National Petroleum Resources

6

Petroleum resources and the management of national wealth

IULIE ASLAKSEN, KJELL ARNE BREKKE, TOR ARNT JOHNSEN AND ASBJØRN AAHEIM

6.1 BACKGROUND

The oil price shock in 1973/74 spurred a new interest in the economics of natural resources. Scarcity of important natural resources was acknowledged as a potential obstacle to continued economic growth. An extensive branch of the literature (see e.g. Dasgupta and Heal, 1979, for a review), is devoted to the question of whether resource scarcity can be overcome by capital substitution. For a resource-exporting country, other important policy questions arise.

One of these is whether revenues from resource exports should be used to increase the level of consumption, for domestic capital formation or to build up foreign assets. In this context, resource depletion becomes part of the overall problem of national wealth management. This is the viewpoint taken for example by Aarrestad (1978, 1979) and Dasgupta, Eastwood and Heal (1978). They analyse optimal resource depletion and capital formation in an open economy. They derive the now familiar conclusion that optimal resource depletion is determined by equalizing the rates of return on three kinds of assets: domestic and foreign capital and natural resources. Their wealth approach underlies our attempt at conceptualizing and measuring resource wealth and its rate of return.

For countries with large natural resources, income from resource extraction is certainly a 'cause' of their wealth. However, there is no way to account for changes in wealth of natural resources which is commonly agreed upon. According to the United Nation's Standard of National Accounts (SNA), the net value of resources extracted is accounted as income, while unextracted resources are not included in the accounts, thus taking no account for changes in the wealth of natural resources. This has brought about criticism of the SNA system. Since petroleum is a non-renewable resource, the size of the remaining reserves will decrease when petroleum is extracted (unless new fields are found). It is quite common to assume that the same applies for the wealth of petroleum. As we shall see, however, this is not necessarily correct.

Several suggestions as to how changes in wealth should be accounted for have been discussed (e.g. Ward, 1982; Stauffer, 1986). In Norway, much emphasis has been put on the correct measurement of national saving. The Norwegian National Accounts show an immense increase in national savings in the first half of the 1980s as a result of the fast increase in petroleum extraction and the oil price shock at the start of the decade. The question then arose as to what the relevant measure of saving is if changes in the wealth of petroleum are taken properly into account.

Skånland (1985) sees petroleum extraction as depletion of wealth only, and obtains 'corrected' savings ratios simply by subtracting the annual resource rent from national income. However, as Strøm (1986) emphasizes, this is depletion of a non-existing wealth if the wealth of petroleum is not included in the National Accounts. Skånland's savings ratio therefore turns out very low.

Strøm (1986) instead calculates savings ratios where changes in the wealth of petroleum reserves are accounted for. He defines petroleum wealth as current net price times reserves. This coincides with the expected net present value if expected future net price of oil and gas increases at a rate equal to the discount rate (the Hotelling rule). Large fluctuations, especially in the price of petroleum, then also cause extreme fluctuations in the savings ratio (from 18 to 60% in the period 1980–84).

In NOU (1988) an attempt to correct for this by calculating actual saving is made: wealth is calculated as if prices and reserves were known in advance. For example, the wealth of petroleum in 1980 is calculated by assuming actual development of prices from 1980 till 1988, and then expected prices from 1988 and beyond. However, these results cannot be used as a background for evaluation of the management of the wealth of petroleum, since they are based on information that was not available when the policy decisions were made. Recommendations based on these calculations will therefore have a strong character of hindsight.

It is also important that policy recommendations drawn from calculations of national wealth corrected for the wealth of petroleum are illuminated in a wealth management context. This is not always the case. Rather, a high savings ratio seems to be regarded as a measure in itself: if the savings ratio corrected for the wealth of petroleum turns out lower than that reported in the National Accounts, there is a tendency to recommend a more restrictive economic policy, and vice versa. So, there is also a need for a discussion of how changes in wealth may change the optimal level of consumption and saving.

Calculations of the wealth of oil and gas

In this chapter, we will try to clarify the concept of 'wealth of petroleum', show how changes occur and discuss how optimal consumption and national saving may be influenced by changes in the wealth of petroleum. The results are illustrated with calculations on Norwegian data. In section 6.2 the wealth of petroleum is defined, and changes in this wealth are discussed in section 6.3. Calculations of the wealth of petroleum and changes in this wealth are presented and discussed in section 6.4. In section 6.5 we present an intertemporal model for optimal management of national wealth. The model is applied in a discussion of the consumption level and the development of total Norwegian national wealth in the period 1973–89.

6.2 CALCULATIONS OF THE WEALTH OF OIL AND GAS

Suppose we have, at time t, a wealth W_0 and get a certain stream of revenues I_t. How much can we spend for consumption? With a perfect capital market, yielding return r, the long-run constraint on consumption is

$$\sum_{s=t}^{\infty} C_s (1+r)^{t-s} = W_0 + \sum_{s=t}^{\infty} I_s (1+r)^{t-s}$$

The present value of future revenues appears in the same way as initial wealth, hence it is naturally included in an extended wealth concept. But suppose that some of the revenues I_t are stochastic, what then is the wealth of these revenues? This is exactly the problem that appears in calculating petroleum wealth. Since future prices of petroleum and future production are uncertain, petroleum revenues are also uncertain. In this chapter we have solved this problem by defining petroleum wealth as the expected present value of future revenues, but as we shall see, this definition is subject to several limitations. For example, it can only be defended for a very restrictive class of preferences. Still the approach is valuable, due to the simplicity of the management of spending wealth compared to spending stochastic income.

Petroleum wealth is the value of the remaining oil and gas fields. The value of previously extracted oil and gas is not included in the wealth of petroleum. However, a part of it is accumulated in other wealth components. Since oil fields are rarely sold in markets, we cannot observe the value directly, but have to estimate the value as expected future income. This income depends above all on the future price of oil and gas.

Consider an oilfield where we can only choose the date for starting the development. Once development is started, the production and cost paths are given. A decision rule for when to start development of the oilfield will give the start time τ as a function of the development in the oil price and other variables that are part of the decision rule. Formally, τ is a stochastic time, where we will know at time t whether or not $\tau \leq t$ (in mathematical literature τ is called a stopping time). An example of such a decision rule is the first time the price reaches a prescribed level.

Define the value of such a field as

$$\max_{\tau} \left(E^t \left[\sum_{s=0}^{S} P_{s+\tau} x_s (1+r)^{\tau-s} \right] \right) \quad (6.1)$$

where E^t denotes the expectation with respect to the probability distribution of future oil prices at time t, x_s is the quantity extracted at s periods after the start time, $P_{s+\tau}$ is the net price at $s + \tau$, and r is the discount rate. S is the expected date of depletion. Since we will concentrate on the wealth management of the decision maker, the relevant probability distribution is the planner's subjective probability distribution.

Applications of this definition are difficult for two reasons:

1. The optimal starting time τ has to be computed as a flexible strategy, i.e. as a function of the oil price to be revealed rather than as a definite time decided at t (constant strategy).
2. The decision maker's subjective probability distribution of future oil prices is unknown.

To the first problem: the calculation of a flexible strategy is difficult. The restriction to constant strategies will lead to solutions that are clearly suboptimal. In Brekke, Gjelsvik and Olsen (1988) the optimal value of the Snorre field of the North Sea is calculated at Nkr9 bn under a flexible strategy, while restricted to a constant strategy, the field has a negligible net value. To consider only constant rather than flexible strategies is more biased when the oil price barely exceeds unit extraction cost. In the early 1980s, however, immediate development was optimal or near optimal for most fields, and since this is a constant strategy, the restriction to constant strategies is no severe restriction in estimating wealth in this period.

To the second problem: it is hard to obtain data from which the subjective probability distributions of the decision makers can be deduced. Governmental White Papers contain information that can be interpreted as the expectations of the development in future oil prices. The price paths are not stated as expectations with respect to any subjective probability distribution, but to interpret them as such may be an allowable short-cut. If we restrict the strategies to constant strategies, i.e. we restrict τ to be any fixed time T, independent of the development in prices and other relevant information, we may use these expectations to estimate wealth.

We have not yet considered the required rate of return in the calculation of the present value. Here we could draw on results from the financial literature. In the capital asset pricing model the required rate of return depends on the covariance between the market portfolio and the rate of return of an asset with the same stochastic properties as the petroleum price. The relevance of the capital asset pricing model on wealth management models is, however, not clear. We will therefore disregard the problem that future investments in petroleum fields are uncertain and should be discounted at a different rate from the oil revenues. We consider only the net price of petroleum, and will therefore apply the discount rate r, which is derived from the two discount rates for the oil price and the extraction cost, respectively. This leads to the following definition of petroleum wealth.

DEFINITION 1. *Let $P_s^t = E^t[P_s]$ be the expected net price in period s viewed from period t. Let (x_s^t) be the path of extraction that is optimal within the set of constant strategies as decoded at t. Then oil wealth at time t is defined as*

$$W_t = \sum_{s=t}^{\infty} P_s^t x_s^t (1 + r)^{t-s} \qquad (6.2)$$

6.3 CHANGES IN WEALTH AND EXTRACTION OF OIL

Any wealth has a positive expected return. This is a consequence of defining wealth as discounted future expected revenues. It means that we require the same rate of return from oil wealth that we can get from other wealth objects. Hence expected return from the wealth is exactly the rate of return required in the discounting. Since oil prices are uncertain, the rate of return on oil wealth is also uncertain. The actual rate of return may thus turn out to be negative. Changes in the wealth of petroleum may thus have different causes, and are not only a result of decisions taken by decision makers.

6.3.1 Certain future oil prices

We will start very simply by assuming that future net oil prices P_t are certain, and grow at a rate equal to the risk-free rate of return r, $P_{t+1} = P_t(1 + r)$, according to the Hotelling rule. Let S_t be remaining reserves at time t, and assume that the extraction path x_t is exogenously given. In this case we have

Oil wealth at time t is $\qquad P_t S_t$

We extract x_t, hence $\qquad S_{t+1} = S_t - x_t$

Oil wealth at time $t + 1$ is $\qquad P_{t+1} S_{t+1} = (S_t - x_t) P_t (1 + r)$

Note that the petroleum wealth is increasing if $(S_t - x_t) P_t (1 + r) > S_t P_t$, or equivalently if

$$x_t < \frac{r}{1 + r} S_t \qquad (6.3)$$

Hence, for x_t sufficiently small oil wealth increases even with positive extraction, and even though no new fields are found.

The assumption that $P_s = P_t(1 + r)^{s-t}$ is not reasonable. We now replace this assumption by the assumption that the price is still deterministic, but follows an arbitrary path. For the moment we will also stick to the assumption that the extraction path is exogenously given.

What is the value in period t of x_s units of oil sold at net price P_s in period s? Suppose we may buy or sell an asset with certain return r. Investing $P_s x_s (1 + r)^{t-s}$ in this asset gives us $P_s x_s$ at time s. The present value of this revenue is thus $P_s x_s (1 + r)^{t-s}$. The wealth defined as present value of future revenues is

$$W_t = \sum_{s=t}^{\infty} P_s x_s (1 + r)^{t-s} \tag{6.4}$$

Note that the oil wealth is defined by requiring a rate of return r and under certainty the wealth will actually give this return, independent of the shape of the price path. The conclusion that oil wealth gives a rate of return r is therefore a tautology.

The oil wealth is increasing if $W_{t+1} > W_t$ or equivalently if

$$P_t x_t < r \left(\sum_{s=t+1}^{\infty} P_s x_s (1 + r)^{t-s} \right) = \sum_{s=t+1}^{\infty} \lambda_s P_s x_s \tag{6.5}$$

where $\lambda_s = r/(1 + r)^{s-t}$ and $\sum_{s=t+1}^{\infty} \lambda_s = 1$. Thus, if the revenue is less than the weighted average future returns, the oil wealth will increase. To put it another way, the oil wealth will increase if the current revenue is less than the return from the remaining wealth. Hence, even if we expect oil prices to fall, and even with positive (but sufficiently small) extraction x_t, the oil wealth may increase.

To illustrate this effect, consider a revenue path over 5 years with Nkr10 bn revenue the first year and Nkr50 bn the next 4 years. The present value of future revenues with 7% discount is Nkr17.9 bn in the first year and Nkr18.1 bn the second year. Hence, the value of remaining revenues increases even with positive revenue the first year.

6.3.2 Uncertain oil prices

So far we have assumed that the extraction path is deterministic and exogenously given. This assumption is more problematic for uncertain oil prices. The crucial part of the assumption is that the optimal path is determinisitic. In general the optimal strategy in the case of uncertainty will be contingent on the stochastic price, and hence the policy will also be stochastic. For fields for which development is approved by the Norwegian Parliament (approved fields) the assumption of deterministic extraction is rather reasonable, since it will rarely be optimal to let production depend on oil prices.[1] For other fields the optimal strategy for start of development will be stochastic. We will, however, stick to the case of a constant extraction strategy as introduced in Definition 1.

What is the value at time t of x_s units of oil sold at the uncertain price P_s at time s? Suppose there exists an asset with return r, with the same risk as the price of petroleum P_s. The equilibrium price at time t for an income at time s with the same probability distribution as P_s is $E^t[P_s](1 + r)^{t-s}$. The definition of oil wealth based on the equilibrium price of future oil production is

$$W_t = \sum_{s=t}^{\infty} E^t[P_s]x_s(1 + r)^{t-s} \tag{6.6}$$

Actual changes in wealth are now:

$$W_{t+1} = W_t = \sum_{s=t+1}^{\infty} E^{t+1}[P_s]x_s(1 + r)^{t+1-s}$$

$$- \sum_{s=t}^{\infty} E^t[P_s]x_s(1 + r)^{t-s}$$

$$= \sum_{s=t+1}^{\infty} (E^{t+1}[P_s] - E^t[P_s])x_s(1 + r)^{t+1-s} \tag{6.7}$$

$$+ r \sum_{s=t+1}^{\infty} E^t[P_s]x_s(1 + r)^{t-s} - P_t x_t$$

The change in wealth is here decomposed in three terms. The first one

$$\sum_{s=t+1}^{\infty} (E^{t+1}[P_s] - E^t[P_s])x_s(1 + r)^{t+1-s}$$

expresses changes in expectation. At time t these changes have expectation 0 since

$$E^t\{E^{t+1}[P_s] - E^t[P_s]\} = 0 \tag{6.8}$$

The second term is the expected return from the wealth remaining after x_t is extracted, i.e. the expected return from

$$\sum_{s=t+1}^{\infty} E^t[P_s]x_s(1 + r)^{t-s}$$

Finally the last term is the revenues from extraction at time t.

As in the case of certainty, expected wealth increases if the revenues from extraction are less than the expected return from the remaining wealth, i.e. if

$$P_t x_t < r \sum_{s=t+1}^{\infty} E^t[P_s]x_s(1 + r)^{t+1-s} \tag{6.9}$$

[1] Due to high maintenance costs on North Sea oilfields, temporary halts are rarely optimal. For a discussion of this see Bjerkholt and Brekke (1988).

110 *Petroleum resources and the management of national wealth*

The two last terms of Equation 6.7, then, express expected changes in wealth, $E^t[W_{t+1}] - W_t$. Decisions taken at time t can only be evaluated on the basis of information available at time t. Note that at time t only expected changes in wealth will be known. Actual changes in wealth include at time t, the unknown term expressing changes in expectations.

6.4 CALCULATIONS

In this section we present calculations of the Norwegian wealth of petroleum. These are based on estimates of physical amounts of resources, future production projections, cost figures of development and price expectations. The figures must be regarded as rough estimates of the wealth.

Total reserves of oil and gas are partly located in fields approved for development by the Norwegian Parliament, and in non-approved fields. Both discovered and non-discovered fields constitute the resources in non-approved fields. Figure 6.1 displays expected reserves divided into approved and non-approved fields for the period 1973–89. The figures are taken from the annual reports of the Oil Directorate (1974–88). The estimates are uncertain, and they are not explicitly related to evaluations of expected extraction costs.

Total resources have been substantially upgraded on two occasions. From

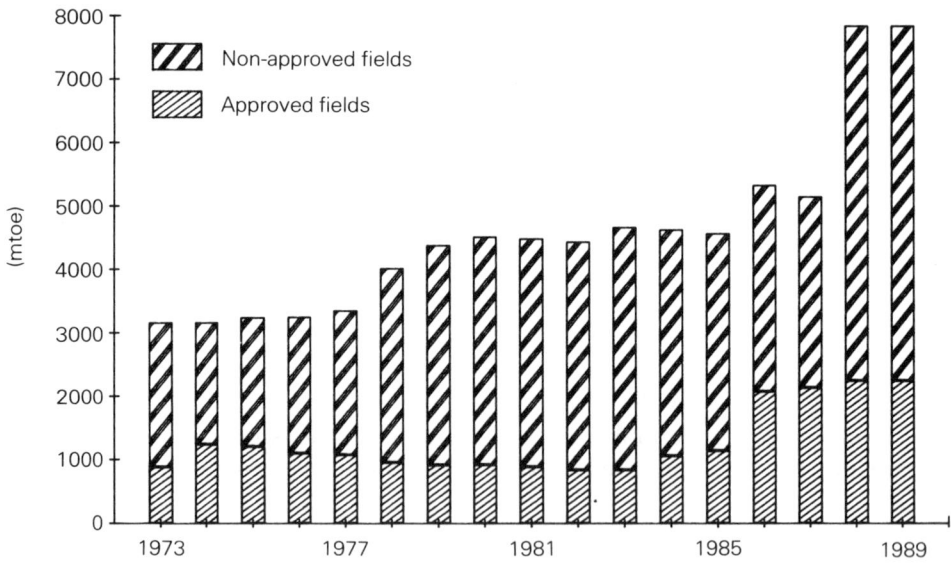

Fig. 6.1 Expected reserves of oil and gas, 1973–89. *Source:* Annual reports of the Oil Directorate (1974–88).

1977 to 1979, upgrading was due to the discoveries of the Oseberg and Gullfaks fields, and to better methods for resource estimation. From 1986 the estimates include northern fields, which has caused an upgrading in later years. The reserves in approved fields increased in 1974 when the Statfjord field was approved. From 1984, the development of several large fields has started, and the 'jump' in 1986 can be explained by the decision to develop the gas fields Troll and Sleipner.

Data on costs and production profiles for all approved fields are taken from Wood and MacKenzie (1986). All capital costs are accounted for the year of procurement. Note that this is not strictly in accordance with the theoretical framework presented in section 6.3. Resources in non-approved fields have been treated as two large fields, one south of the 62nd parallel and one in northern areas. Both fields are assumed to have the same production profile as a typical and recently approved field. The break-even price for the southern field is set to $15 per barrel, and for the northern field the break-even price is assumed to be about $20 per barrel.

From governmental White Papers, especially national budgets and long-term programmes, we have recovered the expectations about the oil-price growth. Figure 6.2 shows the actual oil-price development compared to the expected paths in different years. The gas prices used are collected from Wood and MacKenzie (1986). They are field-specific, but the growth in the gas prices is assumed to be the same for all fields and equal to the growth rates of the oil price, but lagged one year.

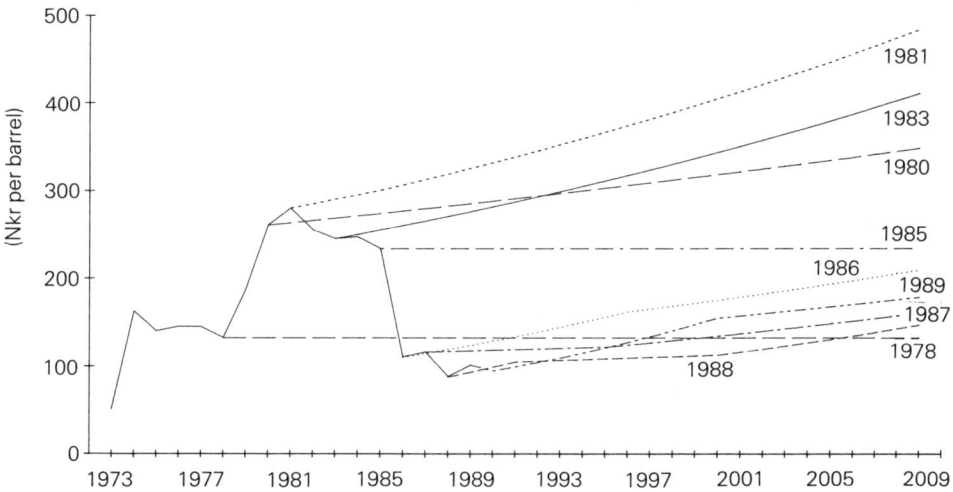

Fig. 6.2 Actual and expected oil price, deflated by the GDP price index of Norway, 1986 = 100. *Source:* Governmental White Papers (national budgets, long-term programmes and others).

112 Petroleum resources and the management of national wealth

The figure shows clearly to what extent the expected level of the future oil price depends on the current price. It also seems to be a strong belief that the future price of oil will increase. To put it bluntly, a normal expectation for the future oil price is a 1–2% increase from the current level, except that the expected increase is adjusted for the change in price level in the latest two years. For two years, 1974 and 1986, the prevailing opinion seemed to be that the current level was, respectively, higher and lower than a 'normal' price.

There are reasons to emphasize that the expected paths in Fig. 6.2 are not official expected paths based on subjective probabilities, but merely 'price projections'. The projection made in one year is therefore not necessarily comparable with the path for another year. For some years, no official projections were found. In these years, the growth rates were taken as the average of the growth rates of the neighbouring years. Furthermore, the White Papers in later years indicate ranges for future oil prices. We have approximated mean values or chosen the path representing the reference alternatives.

Based on the previous assumptions, the expected net present value, or wealth of the reserves in all fields, is calculated according to Equation 6.6. A discount rate of 7% is used. Figure 6.3 displays estimates of the wealth in approved and non-approved fields for 1973–89.

During the 1980s, the wealth of petroleum has fluctuated between Nkr413 and 2273bn. As early as 1983 and 1984, that is 2 years before the collapse in the oil price, wealth was downgraded to large extent as a result

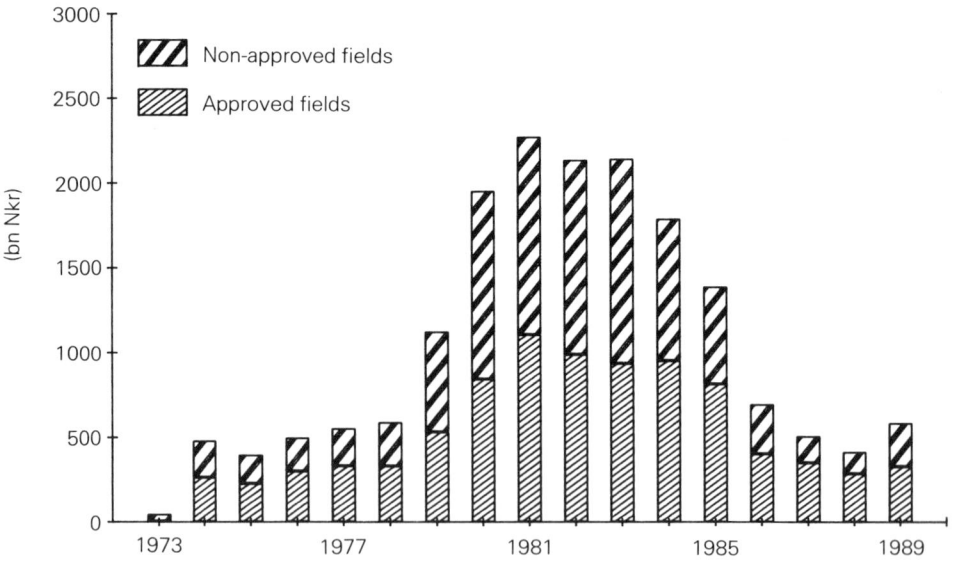

Fig. 6.3 Estimates of total petroleum wealth, 1986 prices.

of a less optimistic view of the development in future oil price. Also in 1987, the expected wealth of petroleum was reduced. This illustrates the lag effect of price changes on expected prices: in 1986, the price level was regarded as extraordinarily low, but in 1987 the new and lower price level had found widespread acceptance.

Table 6.1 shows calculations of the terms explaining changes in wealth according to Equation 6.7, and displays estimates of wealth, changes in expectations, return on remaining wealth and annual revenues from extraction in the period 1973–89. Expectations have fluctuated considerably in this period, mainly because of price fluctuations. From 1981 to 1987, a steady reduction in price expectations caused a reduction in the wealth of petroleum of Nkr2200bn, or Nkr55000 per capita. The positive shift in 1988 is due to re-evaluation of resources, and is highly uncertain. In all years we see that the expected return on remaining wealth has been higher than the revenues. Therefore, one cannot claim that the petroleum wealth has been reduced in this period.

The re-evaluations of the wealth of petroleum will naturally have an impact on the optimal consumption path. For example, when changing expectations causes the wealth of petroleum, and thereby the total national wealth, to be reduced by Nkr2200bn, it will hardly be possible to sustain current level of consumption. The relation between wealth and optimal consumption will be discussed in the next section.

Table 6.1 Petroleum wealth, changes in wealth and revenues 1973–89 (bn Nkr, 1986 prices)

Year	Wealth	Changes in expectations	Expected return on remaining wealth	Petroleum revenues
1973	47	428	4	−4
1974	482	−130	35	−12
1975	398	60	29	−12
1976	499	13	35	−7
1977	554	−6	38	−3
1978	590	504	41	10
1979	1125	777	77	24
1980	1955	239	133	55
1981	2273	−233	155	59
1982	2136	−88	146	51
1983	2143	−441	146	59
1984	1789	−460	121	62
1985	1388	−731	93	56
1986	694	−219	47	16
1987	506	−106	34	21
1988	413	162	27	20
1989	582	—	39	29

6.5 OPTIMAL MANAGEMENT OF NATIONAL WEALTH

The concept of national wealth is meant to represent future consumption possibilities. In a wide sense, all components affecting future consumption possibilities, either directly or via production, are thus a part of the national wealth. These are human capital, real capital and foreign debt, natural resources and environmental resources. In this section we shall confine ourselves to real capital, net foreign reserves and the value of petroleum resources.

Management of the national wealth means making an intertemporal choice between present consumption and wealth formation. This is a question of allocating the national income, generated by production of goods and services and resource extraction, into current consumption and investment for future consumption. To be able to represent future consumption possibilities as wealth, we have to assume that there are no interdependencies between production and consumption and that uncertainty is perfectly correlated with the development in marketable assets. Otherwise, the ideal way to handle uncertainty would be a complete optimization of investment and consumption under uncertain income with all interdependencies explicitly stated. However, such an analysis is very complicated. To solve the model we have to turn to numerical solutions.

A simpler approach is to find an analytical solution to the problem of maximizing an intertemporal utility function, subject to some wealth constraint (the wealth management approach). Although analytical solutions in such models allow only for a narrow class of utility functions, the approach gives an intuition of what the optimal consumption should be. We obtain reasonable rules of thumb, in particular how to treat future uncertain income compared to certain income. The simplicity of this model is its main force.

6.5.1 An intertemporal optimization model

Consider the intertemporal optimization problem consisting of maximizing the expected discounted sum of utility of consumption, $U(C_t)$, over a finite horizon of length $T - t$, taking into consideration the utility of terminal wealth, $V(W_T)$, which is meant to represent the consumption possibilities beyond the planning horizon. The objective function at the beginning of the period, t, is

$$\max_{C_s} E\left\{\sum_{s=t}^{T} U(C_\tau)\theta^{\tau-t} + V(W_T)\theta^{T-t+1}\right\} \quad (6.10)$$

subject to

$$B_t = (1 + r)B_{t-1} + P_t x_t - C_t \quad (6.11)$$

Optimal management of national wealth

where $P_t x_t$ denotes oil revenues and B_t is foreign bonds. Total wealth consists of foreign bonds and petroleum wealth, which will be defined subsequently. We disregard the cost aspect of petroleum wealth and interpret $P_t x_t$ as net oil revenues. The rate of return of foreign bonds, r, is assumed risk-free. $\theta = 1/(1 - \delta)$ and δ is the rate of time preference.

This is a problem of optimization under uncertain income, and an explicit solution requires rather strict assumptions regarding the choice of utility functions. For a more detailed discussion of the solution, see Aslaksen and Bjerkholt (1987). The solution of the model gives optimal consumption and changes in wealth as functions of the wealth of petroleum and other wealth. The expression for the uncertain wealth of petroleum depends on the probability distribution of the oil price.

We will first consider the case of quadratic utility functions. In this case the petroleum wealth is represented by the present value of expected future oil revenues. Hence, the consumption strategy is independent of the parameters of the utility function. It is also independent of the degree of uncertainty.

Next we solve the model with an exponential utility function. In this case the petroleum wealth depends on the parameters of the utility function, and can be interpreted as a certainty equivalent wealth. The certainty equivalent represents the optimal correction for uncertainty and risk aversion: the optimal decision under uncertainty is found by replacing the stochastic variable by its expected value minus a risk premium, depending on the variance and the degree of risk aversion.

Both these assumptions have their shortcomings in terms of empirical relevance; nevertheless, they are important illustrations of the effect of uncertainty.

6.5.2 Quadratic preference functions

We now assume that utility of consumption as well as utility of terminal wealth have quadratic forms. Utility is measured as deviation from the target value, $\bar{C}_t$ and $\bar{W}_T$.

$$U_t(C_t) = -\tfrac{1}{2}(C_t - \bar{C}_t)^2 \qquad (6.12)$$

$$V(W_T) = -\tfrac{1}{2}\gamma(W_T - \bar{W}_T)^2 \qquad (6.13)$$

In order to ensure positive marginal utilities, the conditions $C_t < \bar{C}_t$, $W_T < \bar{W}_T$ must be satisfied.[2] As C_t increases towards $\bar{C}_t$, the risk aversion increases, which means that the willingness of the decision maker to undertake risky projects is lower the higher is the consumption that is already attained. This is regarded as an unattractive feature of the quadratic

---

[2] In a model simulation $\bar{C}_t$ and $\bar{W}_T$ must be so high that C_t and W_T with probability 1 will be below their upper limits.

preference function. It is empirically more reasonable to assume the risk aversion decreases or is constant as higher consumption levels are reached.

(a) Static expectations

We first model the oil price uncertainty as random fluctuations, where the price observed today has no direct influence on our expectation regarding future oil prices. Technically, we assume that the oil price in different periods, $P_{t+1}, \ldots, P_T$, are stochastically independent, and that

$$E(P_s|P_t) = E(P_s) = \pi_s \quad \text{and} \quad \text{var}(P_s) = \tau_s^2.$$

The path of expected oil prices, $\pi_{t+1}, \ldots, \pi_T$ is exogenously given, independent of P_t. (The estimated price expectations presented in Fig. 6.2 are definitely not of this type.)

The optimization problem given by Equation 6.10 can be solved by the technique of dynamic programming. We find the following solution for optimal consumption:

$$C_t = \hat{C}_t + c_t W_{t-1} < \bar{C}_t \tag{6.14}$$

The wealth concept, W_t, thus arising in the solution is given by

$$W_t = (1 + r)B_{t-1} + P_t x_t + NV_t \tag{6.15}$$

where NV_t is the present value of expected future oil production. The interpretation of W_t is thus total wealth including the accrued return to the two components of national wealth, financial assets and oil reserves. With the stochastic specification given above, NV_t is given by

$$NV_t = \sum_{i=t+1}^{T} \frac{\pi_i x_i}{(1 + r)^{i-t}} \tag{6.16}$$

The parameter $\hat{C}_t$ represents the effect on consumption today of the future target levels for consumption and wealth, and it is given by

$$\hat{C}_t = \bar{C}_t - c_t \left\{ \sum_{i=t+1}^{T} \frac{\bar{C}_i}{(1 + r)^{i-t}} + \frac{\bar{W}_T}{(1 + r)^{T-t+1}} \right\} \tag{6.17}$$

The marginal propensity to consume is given by the recursion

$$c_t = \frac{c_{t+1}\theta(1 + r)^2}{1 + c_{t+1}\theta(1 + r)^2}$$

where $c_{T+1} = \gamma$.

As $T \to \infty$ the stationary solution for c_t becomes quite simple. If the rate of time preference equals the risk-free rate of returns, $\delta = r$,

$$c = \lim_{T \to \infty} c_t = \frac{r}{1 + r}$$

It is often assumed that the rate of time preference is less than the risk-free rate of return, $\delta < r$. In this case we find that

$$c = \lim_{T \to \infty} c_t = \frac{\tilde{r}}{1 + \tilde{r}} > \frac{r}{1 + r} \quad \text{where} \quad \tilde{r} = \frac{1 + r}{1 + \delta}$$

Note that the variance of the oil price, τ_t^2, is not a parameter in the optimal solution. Hence the solution is equivalent to the solution with certain oil revenues, with the certainty estimate for the oil price replaced by the expected value. Since a certain income is equivalent to a wealth, we behave as if we own a wealth of size equal to the present value of future oil revenues. The solution for investment in the financial asset is determined residually from Equation 6.11 after C_t is determined.

In the special case where the path of expected future oil prices grows at a rate equal to the risk-free rate of return.

$$\pi_s = P_t(1 + r)^{s-t}$$

we find the familiar result

$$NV_t = P_t S_t \quad \text{where} \quad S_t = \sum_{i=t+1}^{T} x_i$$

which is the valuation formula for oil reserves in the simple case of no uncertainty and zero extraction costs.

(b) Dynamic expectations

We will now consider the case where expectations of future oil prices are proportional to the oil price observed today. A new observation of the oil price thus shift the entire path of expected future prices. Technically, we consider the growth rate of the oil price from one period to the next

$$r_t = \frac{P_t}{P_{t-1}} - 1$$

and assume that the growth rates $r_t, r_{t+1}, \ldots, r_{T+1}$ are stochastically independent, with $E(r_t) = \rho_t$, $\text{var}(r_t) = \sigma_t^2$. The expected oil price in period s, $s > t$, is then given by

$$E(P_s | P_t) = P_t(1 + \rho_{t+1})(1 + \rho_{t+2}) \ldots (1 + \rho_s) \tag{6.18}$$

This assumption seems to be more in line with the estimated price expectations presented in Fig. 6.2.

The solutions for optimal consumption and investment in risk-free assets are as above, only with the expression for the expected present value of future oil revenues replaced by

$$NV_t = P_t \sum_{i=t+1}^{T} \frac{x_i}{(1 + r)^{i-t}} \prod_{j=t+1}^{i} (1 + \rho_j) \tag{6.19}$$

We see that $\rho_i = r$ gives that $NV_t = P_t S_t$. Only in the special case of (i) constant expected growth rate for the oil price and (ii) expected growth rate equal to the risk-free rate of return, will the remaining oil reserves evaluated at the current price represent the present value of future expected revenues.

Both cases of the quadratic model give the conclusion that oil wealth should be calculated in terms of the expected value of future oil revenues. We will now proceed to the exponential model and derive the certainty equivalent, which gives the risk premium that should be deducted in order to take uncertainty into account.

6.5.3 Exponential preference function

The preference functions are given by

$$U(C_t) = B(1 - e^{-\beta C_t}) \qquad (6.20)$$

$$V(W_T) = G(1 - e^{-\gamma W_T}) \qquad (6.21)$$

where β and γ are the coefficients of absolute risk aversion. Constant absolute risk aversion means that the willingness of the decision maker to undertake risky projects is constant, regardless of the consumption level that is already attained. This may be a questionable assumption.

With this preference function we can find the optimal solution for consumption and changes in wealth only in the case of normally distributed oil prices. The expected value is denoted by π_t and the standard deviation by τ_t. The solution for optimal consumption is given in terms of a risk-adjusted present value of future oil revenues, $\widetilde{Y}_t$.

$$C_t = a_t((1 + r)B_{t-1} + P_t x_t + \widetilde{Y}_t + b_t) \qquad (6.22)$$

Here a_t is the marginal propensity to consume, given recursively by

$$a_t = \frac{a_{t+1}(1 + r)}{1 + a_{t+1}(1 + r)} \quad \text{with } a_{T+1} = \frac{\gamma}{\beta}$$

where b_t is a constant term independent of the parameters of the probability distribution. The certainty equivalent wealth, or risk-adjusted present value of future oil revenues, is given by

$$\widetilde{Y}_t = \sum_{i=t+1}^{T} \frac{\widetilde{y}_i}{(1 + r)^{i-t}} \qquad (6.23)$$

with

$$\widetilde{y}_t = \pi_t x_t - \tfrac{1}{2}\beta a_t \tau_t^2 x_t^2 \qquad (6.24)$$

The certainty equivalent wealth is formed by taking the certainty equivalent of the oil revenues in each future period. In this model the certainty equivalent is the expected oil revenues less a correction term proportional to the variance of the oil price and the coefficient of absolute risk aversion. In contrast to the quadratic model, the degree of uncertainty, represented by the variance, directly influences the decision variable.

Note that the correction term in Equation 6.24 depends upon the parameters of the preference function, hence we can only compute the risk-adjusted wealth if we know the decision maker's preferences, and could simultaneously compute optimal consumption. In other words, the risk-adjusted wealth is as subjective as the optimal consumption, and it is necessary to reveal the parameters of the preference functions from observations of policy variables or statements comparing alternative outcomes. This is no trivial task.

Once the certainty equivalent is established, however, it gives a very useful rule of thumb for calibrating certainty estimates of total wealth and optimal consumption. The underlying rationale is that, in the face of uncertainty, consumption should be lower and risk-free wealth accumulation accordingly higher to ensure against future income risk.

6.5.4 Calculations

Risk-adjusted wealth as given by Equation 6.23 is calculated over the period 1974–89. The return from domestic wealth, i.e. real capital, is measured by net national product (NNP) less petroleum revenues, estimated to Nkr386 bn. Since there is no growth in the solution of the model, the return from domestic wealth is referred to a single year, 1986. The level of consumption can therefore not be compared with the actual consumption path in the period 1973–89, but merely gives an illustration of how changing expectations in the wealth of petroleum affects optimal consumption at a given level. The optimal consumption path in the model may be interpreted as the optimal level given NNP at Nkr386 bn and an expected wealth of oil as calculated in section 6.4. Wealth of oil and gas is the only uncertain wealth in the model. The variations in optimal consumption are thus exclusively a result of changes in expected wealth of petroleum. This means that the consumption path follows the estimated expected wealth of petroleum.

Consumption equals the permanent income from the three kinds of assets: domestic and foreign capital and petroleum wealth. r is set at 7%, $\tau = 0.37$ and $\beta = 0.32$. (The calculations are meant to be illustrative, and the parameters are chosen to get a reasonable risk adjustment. $\tau = 0.37$ corresponds with estimated variance of the oil price.) When calculating risk-adjusted wealth of petroleum, only uncertainty in petroleum prices has been taken into account. As we have seen, the uncertainty in estimates of resources may also be considerable. Adjustment for risk may therefore have been underestimated.

Figure 6.4 displays optimal consumption given these assumptions. Again, the re-evaluation of petroleum wealth in the first half of the 1980s should

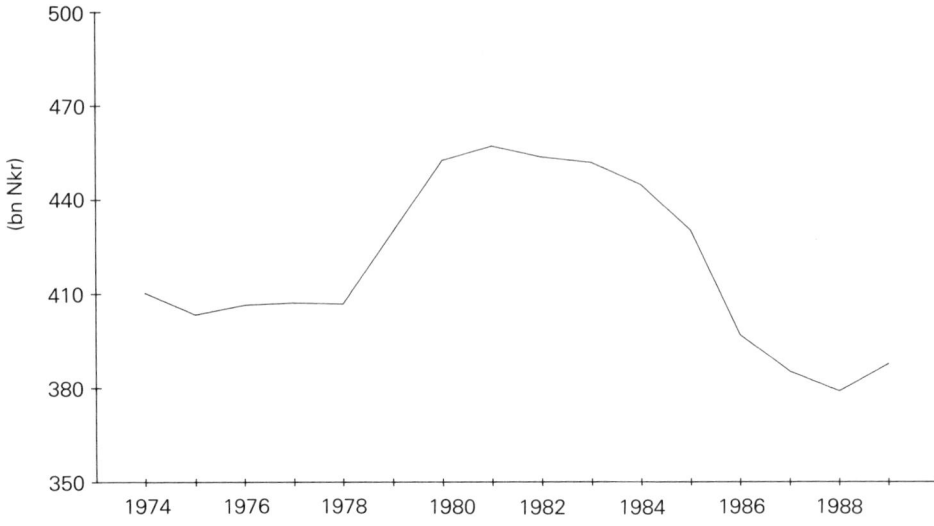

Fig. 6.4 Optimal consumption path 1974–89. Exponential utility function.

restrict consumption according to the model. However, this is partly a consequence of an assumption of time separability in the utility function. A reduction in consumption from 1982 to 1988 at Nkr75bn in itself would probably have an impact on the utility, and thereby on the optimal consumption path.

Figure 6.5 displays net capital outflow according to the National Accounts and the value of extraction less 'permanent income' (the annuity of the wealth). Permanent income is the amount which can be consumed out of the wealth of petroleum each year, rendering the value constant. Since the NNP less income from oil and gas is regarded as certain and held constant at 1986 level in the model, the permanent income less extraction may be interpreted as optimal borrowing to finance current consumption. The increase in national debt will include interest payments on earlier debt in addition.

We first note that the model gives borrowing to finance current consumption in the whole period, even with changing expectations. OPEC I results in a large increase in the wealth of petroleum, allowing for borrowing. According to the model, it would have been optimal to borrow between 1986 Nkr25 and 35 bn per year in foreign markets to finance consumption up till 1985.

The actual development of net flow of capital differs considerably from the optimal path: after a rapid increase in the net inflow until 1977, the inflow turns to outflow in 1980. There are several explanations for these

Optimal management of national wealth

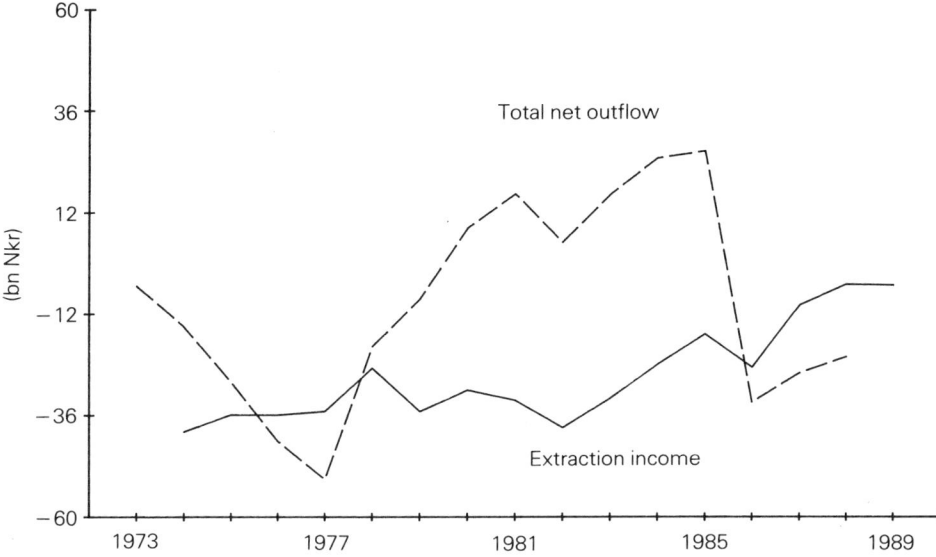

Fig. 6.5 Optimal and actual net capital outflow 1973–89.

differences. First, when considering the actual outflow of capital we must bear in mind that the model treats all other wealth than the wealth of petroleum as certain. In fact, other wealth components may also be uncertain, and the two curves may be different for this reason: in the late 1970s the ocean transport sector went through a difficult period with increasing debts. The prospects for manufacturing industries were also gloomy. Second, a fear of 'the Dutch disease', actualized for example by large capital outflow and high inflation, clearly affected spending decisions in this period. Finally, both wealth and income from oil and gas extraction were increasing in the period. In the model, the 'new' level of consumption is adjusted immediately, due to the assumption of time separability. There are reasons to believe that to the extent that consumption is determined by expected wealth, there is a lag between adjustments in wealth and adjustments in consumption.

In the first half of the 1980s the wealth of petroleum is downgraded, while income from extraction is increasing. The optimal borrowing to finance current consumption decreases steadily. This is also reflected in the actual flow of capital. The change in flow of capital reflects more pessimistic estimates of the wealth of petroleum. However, total consumption in these years increased rapidly, and the common view is that this increase in consumption was too high. Figure 6.5 shows that in spite of increasing consumption, it seems that consumption fitted quite well to the expectations of the wealth of petroleum, and that maybe we could have consumed even more. The difference between the levels of the curves should, however, not be emphasized too strongly, for reasons mentioned above.

In 1986, the fall in the price of oil allows for increased borrowing to finance current consumption. This is because the value of extraction is nearly nil. Even though the expected wealth of petroleum is also reduced, the reduction in current income from petroleum is higher than the necessary reduction in consumption. In 1987, increased value of extraction and decreased expected wealth causes restrictions in the optimal consumption and the optimal amount of borrowing almost diminishes.

The change in 1986 is also reflected in the actual flow of capital, but the shift is more than twice as large as in the optimal solution. This shows the difficulty of adjusting the consumption level in the short term: increasing consumption continued, maybe also reflecting an optimistic view on other national wealth than petroleum. The restrictions in economic policy from 1986 to 1987 seemed, however, to be necessary as a result of re-evaluations in the wealth of petroleum alone. On the other hand, if the difference between the two curves from 1980 to 1985 can be explained by other uncertain factors than the price of oil and gas, there is still some way to go before the consumption level is re-established.

Finally, what happens in the long run? Does the model give borrowing to finance current consumption forever? The answer is, of course, no; we end up with a large positive wealth, larger than the expected wealth today if our expectations today are correct, since we assume risk aversion. But there are no restrictions on the amount of borrowing. Thus, before we start to pay back our debt, the accumulated debt may be very high. This may be impossible for political reasons, and may have an impact on the ability to obtain credit in international financial markets.

ACKNOWLEDGEMENTS

We would like to thank our colleagues in the Research Department of the Central Bureau of Statistics and Steinar Strøm for helpful discussions and comments.

REFERENCES

Aarrestad, J. (1978) Optimal savings and exhaustible resource extraction in an open economy. *Journal of Economic Theory*, 19 163–79.
Aarrestad, J. (1979) Resource extraction, financial transactions and consumption in an open economy. *The Scandinavian Journal of Economics*, 81, 552–65.

References

Aslaksen, I. and Bjerkholt, O. (1987) Optimal oil depletion versus exogenous oil production. A certainty equivalent analysis under risk aversion and uncertain future prices, Memorandum, Department of Economics, University of Oslo, 4 August.

Bjerkholt, O. and Brekke, K. A. (1988) Optimal starting and stopping rules for resource depletion when price is exogenous and stochastic, Discussion Paper no. 40, Central Bureau of Statistics, Oslo.

Brekke, K. B., Gjelsvik, E. and Olsen, Ø. (1988) Utbygging av oljefelter med usikre oljepriser: Eksemplet Snorre. *Økonomiske Analyser*, no. 4, 25–31, Statistisk sentralbyrå, Oslo.

Dasgupta, P. S., Eastwood, P. K. and Heal, G. (1978) Resource management in a trading economy. *Quarterly Journal of Economics*, **92**, 297–306.

Dasgupta, P. S. and Heal, G. (1979) *Economic Theory and Exhaustible Resources*, James Nisbet & Co. Ltd, Digswell Place, Welwyn and the Cambridge University Press.

NOU 1988:21 (1988) *Norsk økonomi i forandring*, Norges offentlige utredninger, Oslo.

Oljedirektoratet (1974–88) *Årsberetning*, Stavanger.

Skånland, H. (1985) Tempoutvalget – to år etter. *Bergen Bank kvartalsskrift*, no. 3, Bergen.

Stauffer, T. R. (1986) Accounting for 'wasting assets': measurements of income and dependency in oil-rentier states. *The Journal of Energy and Development*, **11**(1), 69–93.

Strøm, S. (1986) Oljemilliardene – Pengegalopp til sorg eller glede? *Sosialøkonomen*, no. 1, 21–9.

Ward, M. (1982) *Accounting for the Depletion of Natural Resources in the National Accounts of Developing Economies*, OECD paper, CD/(82)3010, 28 June.

Wood and MacKenzie (1986) *North Sea Report*, London.

7

Oil and gas revenues and the Norwegian economy in retrospect: alternative macroeconomic policies

ÅDNE CAPPELEN AND EYSTEIN GJELSVIK

7.1 INTRODUCTION

When the oil and gas province in the North Sea was discovered, prospects of a wealthy future created optimism in Norway. Engineers, geologists and businessmen prepared for an investment boom, and oil companies lined up for oilfield licences like wasps around a pot of honey. The income expectations led to pressure for social welfare improvements as well as increased private consumption. Environmentalists, on the other hand, warned about the possibilities of accidents and pollution and that fisheries could be harmed. The Ministry of Finance, of course, never celebrates, and in a White Paper (St. meld. no. 25, 1973–74) the Ministry worried about the vulnerability of an economy being dependent upon temporary windfall profits. A moderate extraction policy not exceeding 90 million tonnes oil equivalent (mtoe) annually was recommended and a share of the rent should be saved.

The catchphrase 'Dutch disease' was not yet coined. The analysis in the White Paper contained, however, many elements later to be included in the Dutch disease literature. The theoretical basis for the analysis of the Dutch disease syndrome had, in fact, been elaborated many years earlier in studies of the mineral sector of the Australian economy. A theoretical framework as well as an overview of the Dutch disease literature is given in Corden (1985). Corden defines Dutch disease as the adverse effects on Dutch manufacturing of the gas discoveries of the 1960s, through appreciation of the Dutch real exchange rate. His model includes both direct and indirect effects on other industries of a booming sector. Wijnbergen (1985) uses an

126 Oil and gas revenues and the Norwegian economy in retrospect

intertemporal analysis and disequilibrium models to discuss this kind of problem.

In section 7.2 we briefly review some main conclusions in the Dutch disease literature. We then confront the hypotheses of that literature by describing some macroeconomic indicators for five European countries in order to see how well theory explains economic developments during the 1970s and 1980s. In the ensuing sections we use a macroeconometric model in order to study contrafactual hypotheses. First, we pose the question: What would be the effect on the Norwegian economy if oil had not been discovered? Then we analyse two proposals on how oil revenues should have been managed in order to avoid Dutch disease.

7.2 MACROECONOMIC EFFECTS OF SPENDING WINDFALL PROFITS IN A SMALL OPEN ECONOMY

When analysing the effects of a 'booming' sector on the economy it is useful to distinguish between direct and indirect effects. The demand for capital, skilled labour, engineers and other inputs caused by the oil activity and the direct income to all parties including the government are called the direct effects on the national economy. The effects on the economy from spending these incomes are defined as the indirect effects. Figure 7.1 indicates the indirect effects. The national production possibility frontier is the curve FF, which is also the consumption possibility frontier if the current account is balanced. $U^0 U^0$ is the contour of the utility function indicating the marginal preferences between the two goods, tradables (T) and non-tradables (N)

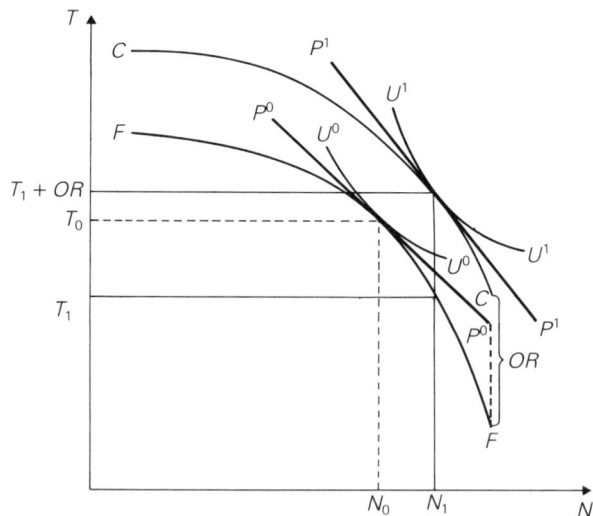

Fig. 7.1 Indirect effects of spending windfall profits.

and the tangent condition leads to the production and consumption T_0N_0. Suppose oil is discovered giving a permanent income addition (OR) measured in units of foreign currency. This does not affect the production possibilities, but the consumption possibility frontier is shifted upwards OR units. If both goods have positive income elasticities, the tangent point is shifted north-east, implying increased consumption of both goods. The production of non-tradables increases to satisfy the increased demand, thus production of tradables has to decrease (but by less than the additional imports OR) to satisfy the increased demand for tradables.

There are three assumptions in this analysis above that need to be discussed. First, the analysis is based on full employment of all resources when oil is discovered. Second, the spending of oil revenues (indirect effects) is assumed not to affect supply of goods or labour (FF curve). The first of these assumptions was satisfied in the Norwegian case, but probably not the second. The spending of oil revenues to expand public employment probably induced a considerable increase in labour supply of women (see section 7.3), shifting the FF curve to the right. Third, the demand shifts require factor flexibility (equilibrium in markets for goods and labour) to allow for the flow of inputs from tradables to non-tradables. The assumption of smooth adjustment to another point on the frontier FF may, of course, be optimistic, at least in the short to medium term, but more realistic in the long term. The shift in relative prices from P_0P_0 to P_1P_1 reflects that prices of non-tradables have to increase, through currency appreciation (appreciation of real exchange rate). The decrease in tradables production and lack of ability to adjust from one long-term equilibrium to another is often referred to as the Dutch disease. If OR is reduced, as was the case for Norway in 1986, balance of payments problems and unemployment may emerge as the economy will have to adjust in the opposite direction.

An important lesson from the model underlying Fig. 7.1 is that to enjoy the benefits of the extra income, you have to spend some of it, but at the expense of production of tradables. Domestic prices will have to rise (real exchange rate effect). These effects are not necessarily signs of illness (Wijnbergen, 1985). However, if a large share of the income is spent, adjustment problems may be difficult to manage in case of sharp income shortfalls like that in 1986. The way the incomes are spent and distributed is also important for economic efficiency (spending affects the supply side) and welfare distribution. We shall return to some of these effects in section 7.3.

7.3 REVIEW OF HISTORICAL DATA 1970–88: AN INTER-COUNTRY COMPARISON

In this section we shall compare the macroeconomic development in some European countries (oil and non-oil countries) to see whether the Dutch

128 Oil and gas revenues and the Norwegian economy in retrospect

disease features can be traced in the data or not. We shall focus on the following issues with emphasis on the Norwegian experience:

1. Can we detect a shift of resources away from the tradable goods sector due to spending of oil revenues?
2. Can we observe higher inflation in countries which have spent oil revenues as prices on non-tradables will rise more rapidly in these countries assuming no change in exchange rates?
3. Has spending of oil revenues resulted in higher consumption (private and public)?
4. If all resources are fully employed when oil is discovered, we should observe a slower rate of growth in non-oil output due to the direct effects of the oil sector. If not, there is indication of positive effects on total supply.
5. As the rate of growth of non-oil GDP is reduced relative to other countries due to direct and indirect effects and inadequate policies towards sunset industries, total GDP may, as a result of increased misallocations, increase less than if oil were not discovered at all. The possibility of such perverse developments has been suggested by some Norwegian economists.

A narrow interpretation of the Dutch disease hypothesis would focus mainly on issue 1. In our analysis, we have chosen a wider approach, and compared the Norwegian development with neighbouring Sweden and Denmark (one a non-oil, non-EC country, the latter an EC member with small oil revenues) in addition to the two EC members, the Netherlands with gas exports and the UK, the biggest petroleum producer and exporter in Europe.

Table 7.1 verifies that the oil and gas shares of national income in the Netherlands and particularly the UK have been lower than in Norway. The Netherlands and the UK have followed a macroeconomic policy different from Norway by reducing government deficits and building up foreign assets through trade surpluses (Byatt *et al.*, 1988). Relative to the national economy, the *OR* shift in Fig. 7.1 has been larger in Norway, smaller in the UK and the Netherlands.

Table 7.1 Oil and gas share of GDP (%)

	Norway	Netherlands	UK
1975	0.0	3.7	n.a.
1980	11.2	5.3	2.8
1985	13.0	8.0	4.3
1986	5.2	5.0	2.0

Some important macroeconomic events in Norway should be noted at the outset:

1. 1970–74: development of the first oilfields;
2. 1973–74: first oil price rise, increasing income expectations;
3. 1974–77: expansive government policy;
4. 1978–80: turn-round in economic policy: devaluation, price and wage freeze;
5. 1980–83: booming oil price expectations, rapidly increasing oil and gas production;
6. 1984–86: deregulation of the credit market, plummeting savings ratio, consumption boom and increasing excess demand;
7. 1986: price collapse in the oil market;
8. 1986–88: new macroeconomic contraction due to increasing deficits. Devaluation, wage freeze. Enhanced contractive effects of increasing savings ratio after the previous 'consumption bonanza'. Increasing unemployment.

The developments of the petroleum sectors in the UK and the Netherlands are quite similar. The Netherlands already had significant gas revenues in the late 1960s, while the first British fields came on stream a couple of years later than in the Norwegian sector. See Byatt *et al.* (1988) for a comparison of policy responses.

(a) GDP

Figure 7.2 shows the volume of GDP of Sweden, Denmark, the Netherlands and the UK in national currencies relative to the Norwegian GDP. A decreasing curve thus implies that GDP has been growing at a slower rate than the Norwegian GDP. Except in 1986–88, after the oil price collapse and the following contraction in Norway, the Norwegian GDP has on average been growing at 3.6% per year, compared to 2.0, 1.7, 1.7 and 2.1% for Denmark, Sweden, the Netherlands and the UK respectively. The Norwegian economy was some 20% bigger relative to the others in 1988 than in 1970. Thus, there is no doubt that total GDP has increased. The Norwegian onshore GDP (oil and gas sectors excluded) grew 2.3% annually, even bigger than the GDP for Sweden and Denmark (Fig. 7.3). Thus, judged by the rapid growth of onshore GDP and the allocation of domestic resources to the oil sector, there are signs of positive supply effects due to the spending of oil revenues.

(b) Private consumption

The relative consumption paths follow similar trends as the GDP for Sweden and Denmark, but the modest gains over Sweden and the UK seem

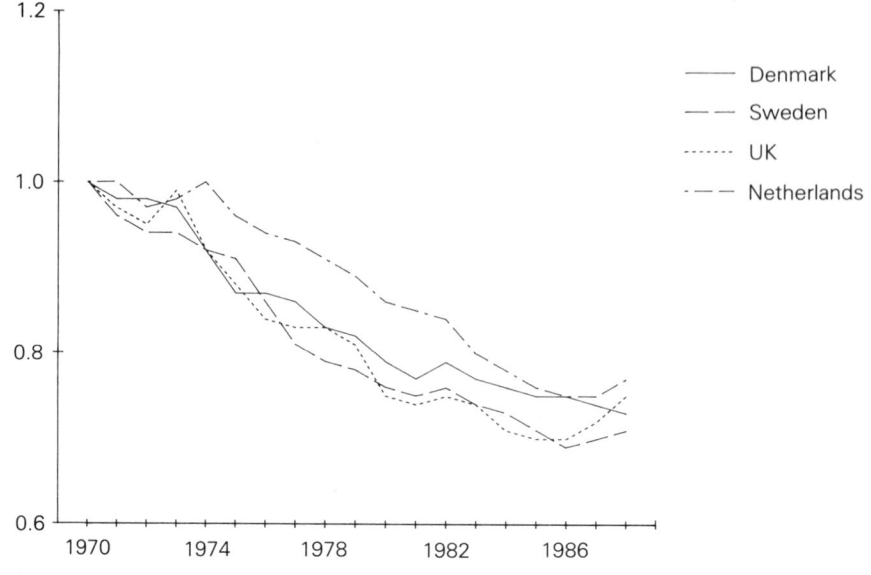

Fig. 7.2 GDP relative to that of Norway.

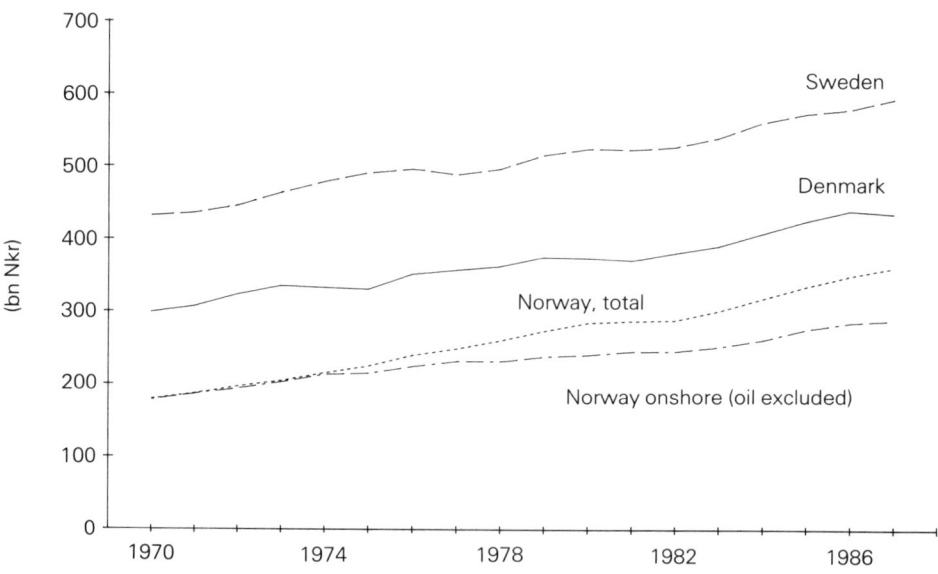

Fig. 7.3 Total and onshore Scandinavian GDP, volume.

to have been lost after 1986. As share of GDP, private consumption in Norway has been decreasing from a low level of about 54% in 1970, to below 50% in the high oil-rent years. Sweden's path is similar to the

Norwegian, while the Dutch and British shares have been stable around 60%. Figure 7.4 suggests that the three countries that have benefited from oil and gas have had a more rapid growth in private consumption. This is, of course, what we should expect.

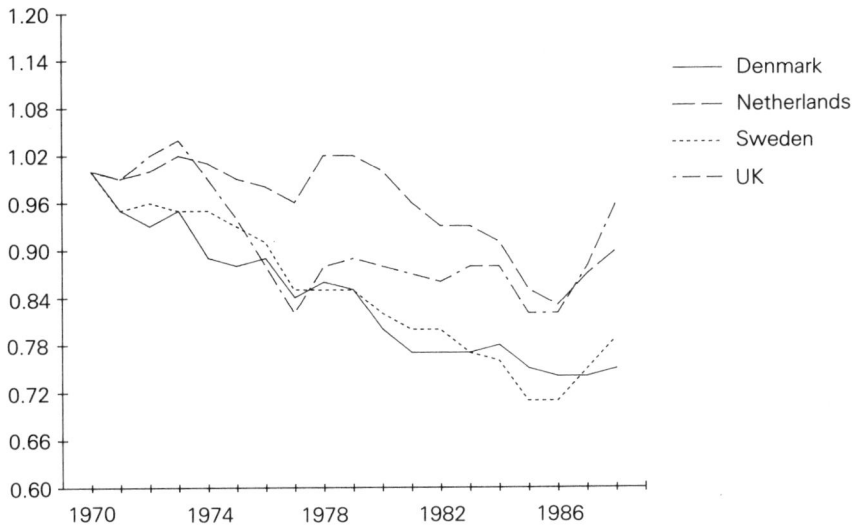

Fig. 7.4 Private consumption relative to that of Norway.

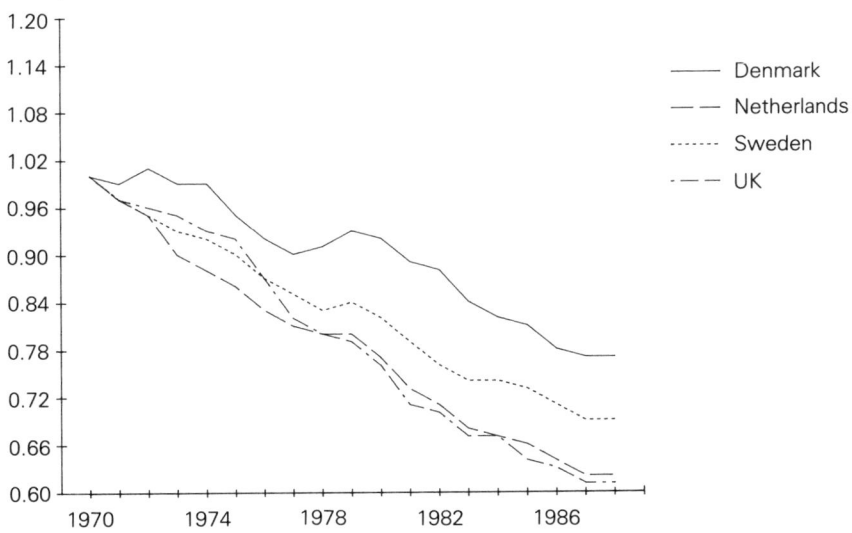

Fig. 7.5 Government consumption relative to that of Norway.

(c) Government consumption

Government consumption is also decreasing in all the countries relative to that of Norway. As share of GDP, government consumption in Norway increased from around 17% to 20% over the period. The sum of government and private consumption share has been stable around 70%, leaving around 30% for investments, but in high-income years with a surplus of current account a lower total consumption share can be observed.

Comparing the trends in Figs. 7.4 and 7.5 the most noticeable difference between Norway, the UK and the Netherlands is that relative growth of public consumption has been higher in Norway. The spending effect in Norway thus seems to have come through different channels than in the other two countries, indicating a difference in political preferences.

(d) Employment

In the mid-1970s the impact of the oil sector on employment became significant. Until 1988 total Norwegian employment increased by about 20% relative to the others (Figure 7.6). In the aftermath of the international recession 1974–75 after the first oil price shock, the Norwegian government pursued a very active macroeconomic policy aimed at counteracting the recession. The trend can also be seen in government employment, but the relative growth is smaller versus the other Scandinavian countries. The unemployment rates (Fig. 7.7) verify a similar trend versus the UK, the Netherlands and Denmark, while Sweden has performed even better than

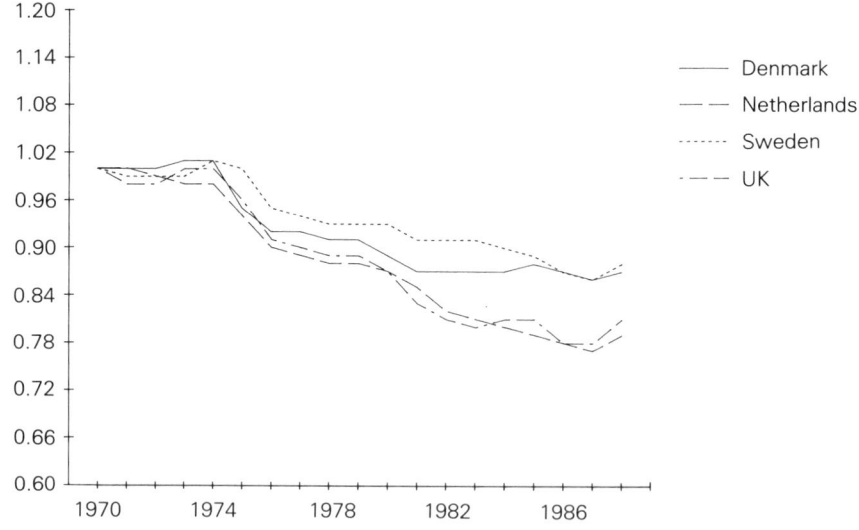

Fig. 7.6 Total employment relative to that of Norway.

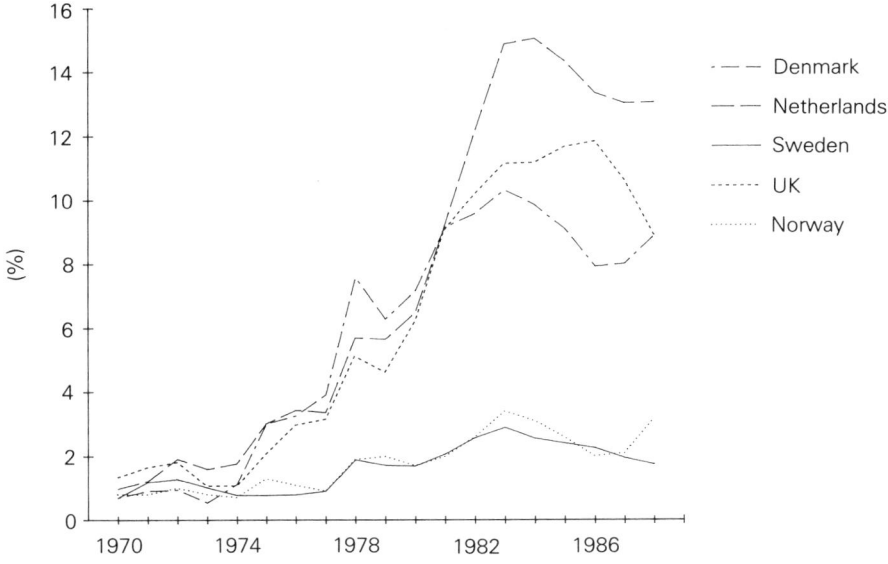

Fig. 7.7 Unemployment rates.

Norway in some periods. Being an even more important political indicator than employment, unemployment rates seem to be more dependent on political preferences than access to oil and gas revenues. However, recently Norwegian unemployment has increased rapidly and is expected to reach 5% in 1989.

As noted above, Norway has used its oil revenues to boost public spending as opposed to the UK and the Netherlands. This may explain why unemployment rates have behaved so differently between these countries. If one regards the Dutch and British unemployment as being mainly 'Classical', spending oil revenues by cutting taxes should leave employment fairly constant, while it is always possible to increase total employment by increasing public activities irrespective of whether unemployment is 'Keynesian' or 'Classical'.

(e) Tradable goods sector indicator

Relative to the Danish, Dutch and Swedish tradable sector, the Norwegian tradable sector has been diminishing. Norway and the UK have on average had similar trends. The curves in Fig. 7.8 thus seem to verify the contractive effect of oil-rent spending on the tradable goods sector. The Netherlands seem to have succeeded better in sheltering manufacturing industry from this effect, so the 'disease' may not be typically Dutch after all! However, in Fig. 7.8 it is most of all the Danish development that conforms well with

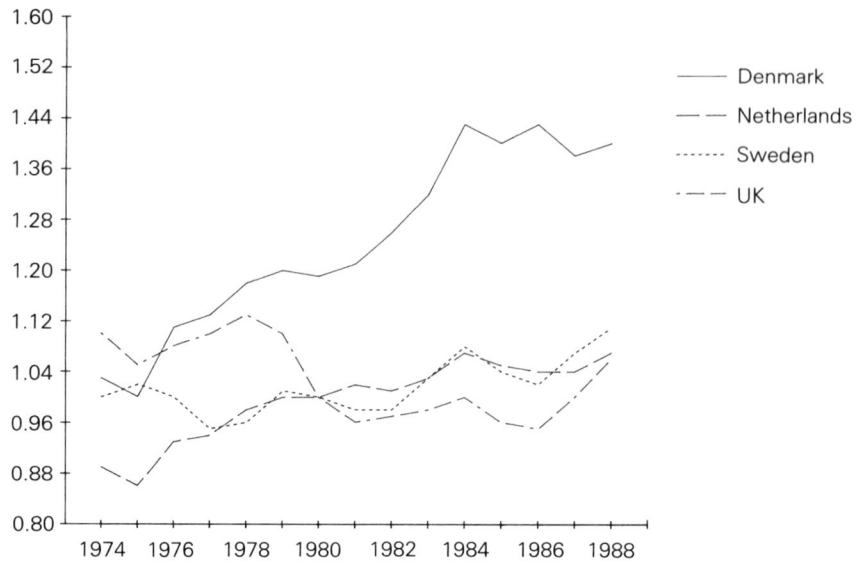

Fig. 7.8 Tradable goods sector indicator. Relative to Norway.

the 'illness hypothesis' with regard to the trading sector. The growth of the Swedish trading sector has not been very different from the Norwegian, and also the UK development has been quite similar to the Norwegian in spite of a relatively smaller oil sector in the UK.

(f) Inflation

Inflation measured by the deflator for consumer expenditure should increase as an indirect effect of oil-rent spending through the additional demand for non-tradables. Compared to Norway's Scandinavian neighbours there is no such trend, except for 1981 and 1986–87. The first peak was an effect of the suppressed inflation being built up during the price freeze of 1979–80, and released during 1980–81. The 1986–87 inflation was mainly a consequence of the 12% devaluation in May 1986, but probably also of excess aggregate demand due to both oil-rent spending and deregulation of credit markets. Compared to the Netherlands and the UK and to some extent Denmark, Norway had high inflation in the 1980s. The EC countries have followed an anti-inflation policy allowing high unemployment, while Sweden and Norway have tried to avoid unemployment and allowed inflation partly as a consequence of exchange rate depreciation. The data seem to verify that choice of macroeconomic policy stance and goals are more important than access to the freedom of policy tools being provided by oil and gas revenues. However, recent experience (1989) may indicate

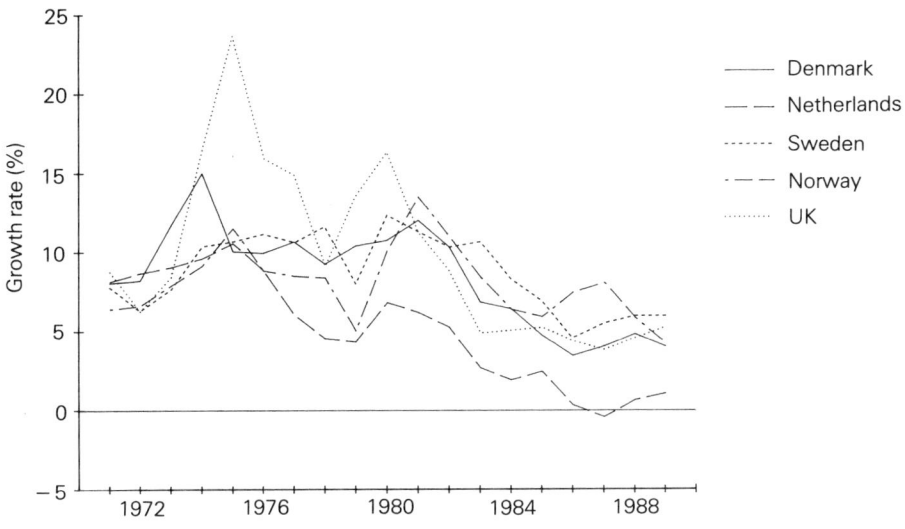

Fig. 7.9 Deflator for consumer expenditure.

that the difference between Norway and the EC countries is not that significant.

(g) A preliminary evaluation

For a long period the oil and gas revenues improved both total and onshore GDP growth of Norway and allowed low unemployment combined with increasing real wages without increased foreign debt. Some expected impacts of spending oil and gas revenues can be traced, but it is hard to single out what is an effect of macroeconomic policy preferences, national institutional differences and traditions on the one hand, and impacts of oil revenue spending on the other. The inter-country comparison indicates that macroeconomic goals and policy are more important. Norway, the Netherlands and the UK have been given extra opportunities, and these opportunities have been handled in various ways. Some events in Norway, however, would hardly have been possible without oil and gas revenues, like the 'bridging policy' and income guarantees to farmers in the 1970s and perhaps even the consumption boom of 1985–86. The relative decrease of the tradables sector was also inevitable in an initially fully employed economy. Rather than maximizing consumption, employment has been promoted through government spending and subsidies. The married women's employment rate has doubled, and the transformation of non-compensated household work to paid work in the public sector has contributed significantly to the GDP growth. Thus, the negative effect on the tradable sector due to spending effects has been reduced by increased supply of labour. In the Norwegian case one can also observe an inflation effect, mainly due to the attempt to assist the tradable sector by devaluations.

7.4 THE IMPACT OF THE OIL SECTOR ON THE NORWEGIAN ECONOMY

In this section we will study some of the macroeconomic effects of the oil sector on the Norwegian economy, using a small macroeconometric model called AMEN. By undertaking historical simulations we will analyse economic developments with and without oil, and we will also study alternatives for the spending of the incomes from the oil sector.

7.4.1 A short description of the AMEN model

This is a small annual macroeconometric model of the Norwegian economy. It was designed for a study very similar to the one we undertake here (cf. Cappelen, Offerdal and Strøm, 1985). Theoretically the model is similar in structure to the larger CBS models KVARTS and MODAG; AMEN disaggregates the Norwegian economy into four sectors, the oil sector, shipping, government and all other sectors called the private mainland sector. The latter sector is modelled econometrically, while the other three enter the model only through accounting identities and simple 'bridging' equations. The parameters of the model are estimated using National Accounts data for 1962–1985. A description of the most important equations is given in Cappelen, Offerdal and Strøm (1985). For the present analysis, the model is re-estimated using more recent data and a new wage equation is included.

The main shortcomings of the model are the lack of financial sector and equations determining nominal interest rates and the exchange rate. The reason why these variables are exogenous in AMEN is partly that the financial sector was highly regulated until some time in the mid-1980s. However, from 1984 onwards financial markets have been deregulated.

We do not intend to give a complete presentation of AMEN in this chapter. The main characteristics of the model will hopefully become clear when we interpret the simulation results later. A summary of the model's tracking performance during 1967–85 which corresponds to the estimation period, is given in Table 7.2 presenting relative root mean square errors (RRMSE) for some main aggregates from a full dynamic simulation starting in 1967. It is not easy to evaluate the significance of the figures in Table 7.2 by themselves. Comparing the RRMSEs with results from similar studies carried out on the CBS quarterly model KVARTS indicates that AMEN

The impact of the oil sector on the Norwegian economy 137

Table 7.2 Historical tracking performance, 1967–85

	RRMSE (%)
GDP	1.8
Private consumption	2.6
Consumer prices	1.3
Hourly wage rate	2.6
Labour supply	0.5
Labour demand	0.5
Exports (mainland)	5.4
Gross investment (mainland)	9.9
Net foreign debt	5.8

performs quite well considering the length of the simulation period and the aggregate nature of the model.

Figures 7.10–7.13 present graphically historical and simulated values for GDP, consumer prices, employment and the current account. We believe this evidence gives at least some support to the usefulness of the model as a tool for counterfactual studies. We would also like to emphasize that the only dummies included are one in the consumption function which captures the switch in purchases of consumer durables between 1969 and 1970 when VAT was introduced, and one in the wage equation to capture the wage freeze from 13 September 1978 to 1 January 1980. There are however, signs of underprediction from 1985 to 1988 due mainly to the deregulation of financial markets which created a boom in domestic demand. The model is not able to track this development closely although it is not off track in 1988. In our simulations we will have to rely on the marginal properties of the model even after 1984.

7.4.2 Norway without oil

This section is an update of parts of the study in Cappelen, Offerdal and Strøm (1985). We simply assume that oil and gas was never discovered, i.e. that no exploration, investment and production ever took place. In the actual historical tracks exploration started in 1965 but the impact of the oil sector on the Norwegian economy was not noticeable until the early 1970s when production started (1971) and investments became substantial. Even though our calculations start in 1967 we report results only from 1971 to 1988.

The analysis is carried out assuming no change in economic policies compared to what actually happened. This was called a passive approach in our 1985 paper and shows therefore the direct effects of the oil sector on the Norwegian economy while the indirect or spending effects are kept

Fig. 7.10 GDP. Simulated minus historical value.

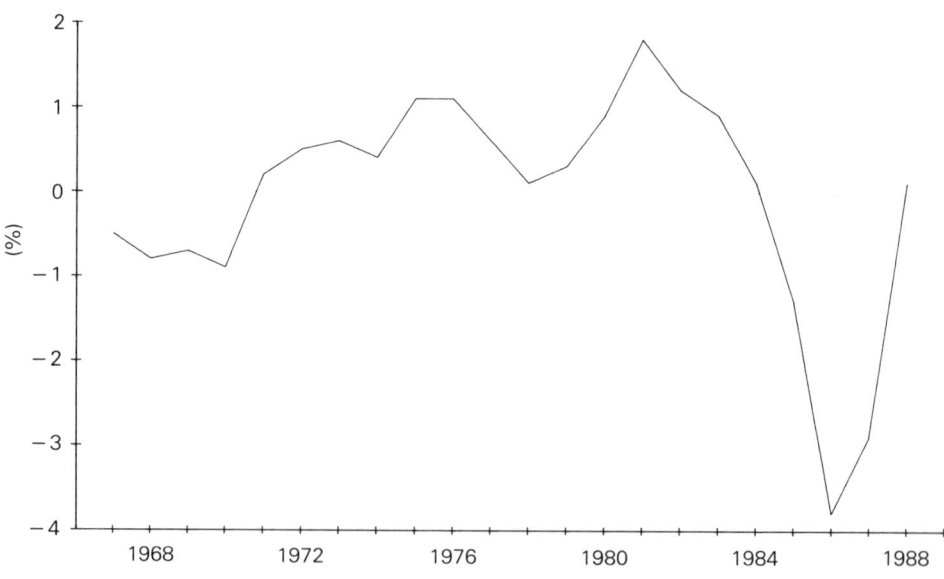

Fig. 7.11 Employment. Simulated minus historical value.

Fig. 7.12 Consumer prices. Simulated minus historical value.

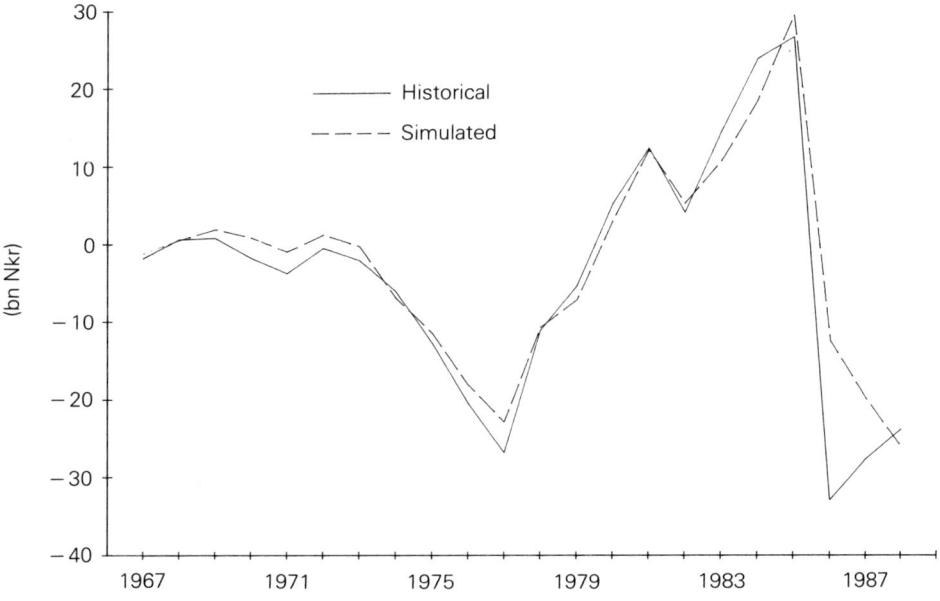

Fig. 7.13 Current account.

Table 7.3 Percentage deviation from historical values due to the elimination of the oil sector. No policy adjustment

	GDP (1980 prices)	GDP private mainland sector	Total employment	Mainland exports	Imports[a]	Private consumption
1971	−0.4	−0.3	−0.1	0.1	−2.3	−0.3
1972	−1.4	−0.5	−0.1	0.2	−3.9	−0.5
1973	−1.6	−0.9	−0.1	0.4	−8.0	−0.8
1974	−2.3	−1.5	−0.2	0.5	−12.5	−1.3
1975	−5.5	−1.4	−0.3	0.9	−14.2	−1.8
1976	−7.5	−1.5	−0.2	1.4	−19.3	−2.1
1977	−8.0	−1.3	−0.1	1.8	−21.0	−2.1
1978	−11.6	−0.5	0.1	2.0	−13.5	−1.7
1979	−13.7	−0.8	0.1	1.9	−14.7	−1.7
1980	−15.7	−0.5	0.1	1.6	−12.0	−1.5
1981	−14.8	−0.2	0.0	1.3	−10.0	−1.3
1982	−14.8	−0.3	−0.1	1.0	−10.0	−1.4
1983	−17.2	−0.7	−0.2	0.8	−15.2	−1.8
1984	−18.8	−0.5	−0.4	0.8	−17.0	−1.8
1985	−19.0	−0.4	−0.5	0.9	−16.1	−1.8
1986	−19.3	−0.1	−0.4	0.9	−14.3	−1.6
1987	−20.6	0.0	−0.4	0.9	−12.4	−1.6
1988	−21.7	0.3	−0.4	0.9	−9.0	−1.7

[a] Excluding shipping services.

constant. Table 7.3 presents percentage deviations for some main economic aggregates between historical developments and the counterfactual path called the passive approach.

The difference between columns one and two in Table 7.3 shows roughly the importance of oil and gas production for GDP. According to our calculations, GDP exclusive of the oil sector, shipping and government sectors would not have been much different without the oil sector. The reason is mainly that by removing the oil investments we also reduce demand both towards exposed and sheltered sectors of the Norwegian economy. Thus, even if the spending of oil revenues may harm the tradable goods sector in the long run, the direct effect of oil investment represents a sustained demand shift in favour of the tradables sector. Lower demand and less employment reduce wages and household demand (private consumption and investment in housing). After some time lower wage rates increase labour demand in spite of lower output and the loss of jobs in the oil sector. This is due to substitution effects in the production function. However, in the long run the effect of loss of jobs in the oil sector dominates. Lower wage rates improve competitiveness and thus increase exports and lower imports. Imports are heavily reduced by eliminating oil investment. Thus initially the current account is improved compared to historical developments (cf. Table 7.4). Just as we found in our 1985 paper, the current account would have shown lower deficit all through the 1970s if oil had not been discovered. Only after OPEC II are the oil price and the oil and gas production high enough to offset the effects on the current account of oil investments. However, as revealed by Table 7.4, the changes in the current account are very large during 1980–88. Turning to net foreign debt as a share of nominal GDP, the ratio is lower until 1980 but increases dramatically thereafter. Obviously, these debt figures indicate that some time around 1980, if not earlier, economic policies would have been very different from what is assumed in this passive approach. That was one of the main issues raised in Cappelen, Offerdal and Strøm (1985) where the simulations ended in 1983 with results very close to those presented in Tables 7.3 and 7.4. The conclusions in that paper with regard to the effects of possible policy adjustment were as follows:

1. A restrictive fiscal policy aimed at reducing the current account deficit would have had severe negative effects on the labour market in terms of higher unemployment.
2. By relying on income policy measures to curb wage inflation and avoiding appreciation of the Norwegian krone from 1973 to 1975, one could have obtained a similar development of the current account as with a restrictive fiscal policy but with much less unemployment.

The latter conclusion of course begs the question of how to implement such an incomes policy. On this question we cannot add much except pointing to

Table 7.4 The effect on the current account and net foreign debt due to the elimination of the oil sector. No policy adjustment

	Historical figures		Norway without oil	
	Current account (bn Nkr)	Net foreign debt (% of GDP)	Current account (bn Nkr)	Net foreign debt (% of GDP)
1971	−3.7	14.1	−2.9	12.2
1972	−0.4	13.8	0.9	10.8
1973	−2.0	12.7	1.5	6.8
1974	−6.2	15.2	1.1	4.4
1975	−12.7	23.7	−6.4	10.4
1976	−20.4	31.9	−11.1	15.6
1977	−26.8	44.1	−16.8	25.2
1978	−11.0	46.5	−12.1	31.2
1979	−5.3	43.4	−13.1	34.1
1980	5.4	32.6	−23.7	37.4
1981	12.5	26.7	−29.1	45.5
1982	4.1	26.9	−43.1	59.6
1983	14.6	22.4	−40.9	70.1
1984	23.9	16.7	−50.6	81.7
1985	26.7	9.7	−56.0	89.3
1986	−32.9	16.0	−84.0	101.1
1987	−27.6	17.6	−85.7	107.6
1988	−23.8	21.2	−84.7	114.8

fairly successful examples of incomes policies in Norway both during 1978 and 1979 and more recently from 1988 and onwards. However, the reduction in the real wage that was calculated in Cappelen, Offerdal and Strøm (1985) is much larger than what actually took place during the two episodes referred to above.

7.5 ALTERNATIVE PROPOSALS FOR HOW MUCH OIL REVENUE TO SPEND

It is impossible to give a definitive answer to the question of how much oil revenue Norway actually spent from the early 1970s to 1988. As was shown in section 7.4, due to the oil sector Norway had larger current account deficits during the 1970s than would have been the case without oil. On the other hand, parts of the deficit came as a result of spending expected but not accrued oil revenues. We start this section with a review of Norwegian oil income policies in the 1980s and refer to two proposals for the management of the oil revenues. Even though this is a normative

Alternative proposals for how much oil revenue to spend 143

question, we do not address the question of what would have constituted an optimal policy and comparing this with the two proposals. A policy that is optimal with regard to saving and consumption would of course have to take the whole economy into account, not only the oil rent. We simply study the positive question of what the effects of alternative spending rules would have been.

7.5.1 Petroleum policy in the 1980s

A number of papers published in Norway during the 1980s have focused on the vulnerability of an economy highly dependent on oil revenues. A report by a government-appointed committee on the future role of the oil and gas sector, NOU (1983: 27), recommended the setting up of an oil fund to separate the harvesting of oil and gas revenues from spending. The report argued, e.g. 'In the first place, it seems more reasonable (than directly considering measures of physical activity levels) to consider both direct and indirect impacts of the oil activity, and then by backward reasoning conclude on activity levels' (p. 14).

> The committee concludes by advising that, for long term planning purposes, one should consider the size of government income relative to the size of the total onshore economy. To fulfil the aim of stable supplies of goods and services from Norwegian industries, the committee recommends regulating the total demand from the petroleum sector. If found appropriate, a minimum demand level can be established in order to maintain and develop skills in supplying industries. Accordingly, a maximum level can be established to avoid a strong dependency of a single sector in the economy (p. 14).

Given the planned production profile and exogenous prices, the petroleum tax revenues will follow, hence a fund is needed to separate spending from revenues, in order to achieve the planned level of 'petroleum dependency', the committee concludes.

These proposals stirred considerable debate between those arguing against any attempt to 'hide away' oil revenues on one side, and those in favour of the committee (mostly representatives from industry branches harmed by indirect effects). The fund was not proposed by the government at the time. However, the idea of an oil fund survived in planning documents, and has been included in the 1990–93 government long-term programme. Even though a considerable share of government revenues has been saved, through amortization of loans and increasing liabilities in central bank government, spending increased and became increasingly expansive in 1984–85.

The other parts of the NOU (1983: 27) recommendations seem to have suffered a similar fate as the fund proposal. Even after 1986 the applications for field developments peaked, with a new demand boom in supplying industries as a result. This would be incompatible with the 'Norwegianization policy' and even might probably provoke OPEC to press for supply restrictions. The government thus proposed a 'development queue' in order to keep the investment level at an 'illustrative level' of Nkr25bn annually, in the spirit of NOU (1983: 27). The level, however, turned out somewhat more illustrative than representative, and is expected to reach Nkr35bn in the early 1990s. Production levels of oil and gas already exceeding 100 mtoe are expected to reach 130 mtoe within 1995. Compared to the NOU (1983: 27, pp. 109–14) recommended/illustrative levels of one path constant around 50 mtoe and a high path of 66 mtoe in 1989, increasing to 86 mtoe in 1995, one can hardly say that NOU (1983: 27) has been a guideline for Norwegian petroleum policy.

In the aftermath of the 1986 turn-round, another committee was established to review the economic prospects of Norway. The report delivered, NOU (1988: 21), recommended that the wealth balances of oil revenue spending should be taken into account, suggesting that spending adjusted for wealth accounting had been too high: 'Savings data in the national accounts overestimate wealth increases after 1980, because decreases of petroleum wealth are not accounted for (p. 13). 'Depletion of petroleum reserves in addition to large foreign deficits, implies that one does not care sufficiently for future generations. The level of consumption today seems not to be resulting from a well-founded consideration of the tradeoff between prevailing and future consumption' (p. 13).

The committee chairman argued for a spending schedule based on permanent income, defined as the real return on the oil wealth. This preserves the oil wealth and gives each generation an equal share of the wealth. 'Total savings were strongly overestimated (1980–86). Whether or not the politicians would have followed a less expansive economic policy provided they had more correct information on national income and saving in these years, is a question we shall probably never know the answer to' (Steigum, 1989). The reference here is probably to the computations in NOU (1988: 21), in which the 'petroleum-adjusted' savings ratio is 2–9% lower than the ordinary savings ratio for the year 1981–85. However, in a recent paper, Aslaksen *et al.* (1990) claims that the conclusion is quite opposite. In all previous years, the spending of oil revenues have been lower than permanent income from the offshore sector. The reason for the divergence is that NOU (1988: 21) used historical oil and gas prices to compute the value of the wealth in previous years. These were much lower than expected values at that point in time, which should be used when evaluating historical policies *ex post*. We agree with Aslaksen *et al.* that previous governments cannot be accused of having had wrong expectations, especially not by economists. Thus an *ex ante* 'petroleum-adjusted' savings

Alternative proposals for how much oil revenue to spend 145

ratio would have been adjusted upwards, and lead to increased spending.

We will now take a closer look at the two spending rules presented above.

7.5.2 Spending returns of an oil fund

Let us construct a completely different development of the Norwegian oil sector from what has actually been experienced. Assume that the oil sector has been organized so that only foreign oil companies were allowed to exploit oil and that they were not permitted to use Norwegian resources in the exploitation. Consequently, no direct effects on the Norwegian economy would occur. The companies would be paid a real rate of return on invested capital of 7% in addition to current costs. The residual income – called the oil rent – should be taxed 100% by the Norwegian central government. The oil rent should then have been invested in financial assets abroad and the real return on these assets could then have been consumed without affecting the real value of the fund. The calculated oil rent from 1973 to 1988 is given in the first column of Table 7.5 and the return on accumulated oil rent is given in column 3 of Table 7.5. The oil rent is negative in the early 1970s when production is low, but invested capital – earning a real return of 7% – is already substantial. The oil fund is not of any significance until 1979. Notice that the oil fund increases both due to positive oil rents and due to inflation, i.e. the nominal part of the interest rate is added to the fund each year.

Table 7.5 Oil rent and 'oil fund', 1973–88 (bn Nkr, current prices)

	Oil rent	'Oil fund'	Real return on 'oil fund'
1973	−0.6	−1.7	−0.0
1974	−0.9	−2.8	−0.0
1975	1.5	−1.6	−0.0
1976	3.3	1.6	0.0
1977	2.8	4.5	0.0
1978	7.0	11.8	0.1
1979	13.7	26.1	0.5
1980	31.8	59.5	1.3
1981	36.5	99.9	4.5
1982	37.8	142.9	7.0
1983	48.0	195.5	11.1
1984	59.5	260.4	16.4
1985	63.3	331.0	18.2
1986	17.8	358.4	14.0
1987	13.4	380.4	16.7
1988	3.1	394.2	15.8

In Table 7.6 the estimated cash flow from the oil sector which is defined as production less material input, wage cost and investment is presented. The historical values for the current account minus the cash flow from oil and interest payment on the cash flow give an estimate of mainland current account. By adding financial return 4 on the accumulated oil fund we arrive at column 3, while column 4 is an estimate of a 'warranted' current account deficit based on a constant debt ratio to GDP (mainland). Thus, by comparing the figures in the fourth and fifth columns of Table 7.6 we may say that from 1973 to 1988 Norway has spent around Nkr210bn of a potential oil fund of Nkr394bn.

In order to investigate the possible outcome for the Norwegian economy of spending only the returns from an 'oil fund' we have assumed that a more restrictive fiscal policy would have been implemented from 1975 onwards. We have, however, not taken the differences in the current account in columns 3 and 4 in Table 7.6 too literally, but instead assumed that over the period 1975–88 the accumulated effect on the current account should be in accordance with that difference in order to achieve the same net foreign debt by 1988.

The basic idea behind accumulating a fund and spending only the returns is that spending is deferred until the return is there, rather than looking at

Table 7.6 Mainland current account, 1973–88 (bn Nkr, current prices)

	Observed current account	Cash flow oil sector	Mainland current account	Mainland current account plus return on oil fund	Warranted current account deficit
1973	−2.0	−3.1	1.1	1.1	−2.0
1974	−6.2	−5.6	−0.6	−0.6	−2.5
1975	−12.7	−4.1	−8.6	−8.6	−2.5
1976	−20.4	−5.5	−14.9	−14.9	−3.0
1977	−26.8	−5.5	−21.3	−21.3	−3.0
1978	−11.0	6.5	−17.5	−17.4	−3.0
1979	−5.3	11.3	−16.6	−16.1	−3.5
1980	5.4	31.7	−26.3	−25.0	−4.0
1981	12.5	40.9	−28.4	−23.9	−4.5
1982	4.1	41.0	−36.9	−29.9	−4.5
1983	14.6	41.0	−26.4	−15.3	−4.5
1984	23.9	51.8	−27.9	−11.5	−5.0
1985	26.7	57.1	−30.4	−12.2	−6.0
1986	−33.4	16.2	−49.6	−35.6	−7.0
1987	−27.6	17.7	−45.3	−28.6	−7.5
1988	−23.8	10.8	−34.6	−18.8	−8.0

Table 7.7 Macroeconomic effects of spending the returns of an oil fund compared to historical developments. Difference (%), 1975–88

	Public consumption	Private consumption	Private investment	Exports mainland	Imports	GDP	Employment	Consumer prices
1975	−1.9	−1.6	0.0	0.0	−1.3	−0.1	−0.2	−0.0
1976	−3.7	−3.9	−1.8	0.3	−2.9	−2.2	−1.2	−0.3
1977	−6.6	−7.3	−4.9	1.2	−5.9	−4.1	−2.7	−0.9
1978	−10.3	−10.7	−10.3	2.4	−10.7	−6.4	−4.7	−1.4
1979	−11.8	−8.9	−19.3	3.5	−11.3	−6.1	−5.7	−2.1
1980	−14.2	−9.8	−24.5	4.5	−13.1	−6.7	−6.5	−2.7
1981	−12.9	−10.2	−25.6	5.5	−12.9	−6.2	−6.3	−3.7
1982	−5.6	−8.7	−20.1	6.9	−9.9	−3.9	−3.1	−4.5
1983	−0.5	−6.9	−10.1	7.5	−6.5	−1.9	0.5	−4.7
1984	0.9	−3.9	−1.3	6.6	−2.9	−0.4	2.6	−4.1
1985	−0.1	−1.6	7.2	4.3	0.1	0.4	2.7	−3.2
1986	−1.9	−0.2	11.2	1.8	1.4	−0.3	1.9	−2.3
1987	−3.8	−0.8	11.4	0.2	1.0	−0.5	0.6	−1.6
1988	−0.2	−0.6	7.6	−0.4	1.1	−0.1	0.5	−1.3

possible future income and increasing spending right away. Since the Norwegian government collects most of the oil rent we have assumed lower public spending for consumption and investment purposes in addition to a moderate tax increase in 1975 (as opposed to an actual reduction).

As evident from Table 7.7, public consumption has to be reduced quite significantly in order to obtain the current account target in Table 7.6. This demand cut works its way through traditional multipliers and accelerator effects. Lower domestic demand reduces imports, output and employment. Higher unemployment in the late 1970s reduces nominal wage growth, improves competitiveness and increases exports. What is perhaps the most important feature of the simulations is the fact that if this 'austerity' policy had been followed from 1975 to 1980, economic performance would have improved during the 1980s. However, that would have meant accepting unemployment rates between 5 and 6% from 1978 to 1981. This is actually the level Norway is experiencing in 1989 due to the fact that the adjustment of spending to uncertain oil revenues was postponed until after 1986. In Table 7.8 we show the difference between historical and simulated values for the current account, the 'debt burden' and unemployment in some years. The figures show that the situation in 1988 would have been quite different in terms of the ratio of foreign debt to nominal GDP. Otherwise it is worth noting that the differences in this year are modest. This is also the case judged by the figures in Table 7.7. It is also interesting to note that the very high debt figures Norway had in the late 1970s would have been much the same even in the oil fund alternative. The reasons are high investment in oil and shipping and the international recession.

Table 7.8 Effects on current account, debt ratio and unemployment. Historical and simulated values

	Current account		Debt/nominal GDP		Unemployment	
	History	Oil fund	History	Oil fund	History	Oil fund
1978	−11.0	−0.7	46.5	39.5	1.8	5.1
1981	12.5	35.0	26.7	4.7	2.1	5.3
1988	−23.8	−9.2	20.1	−18.7	3.2	3.0

7.5.3 Spending permanent income from oil wealth

The spending rule now is based on calculations of the petroleum wealth. The wealth figures are taken from Aslaksen *et al.* (1990) and are based on a real rate of return of 7% in order to calculate permanent income. The figures are given in Table 7.9 where the calculated mainland current account is the same as in Table 7.6. Judging from the figures in this table

Alternative proposals for how much oil revenue to spend 149

Table 7.9 Petroleum wealth, permanent income and current account (bn Nkr)

	Mainland current account	Petroleum wealth in 1986 prices	Permanent income in current prices	Mainland current account plus permanent oil income
1973	1.1	47	1.1	2.2
1974	−0.6	481	11.8	11.2
1975	−8.6	398	11.2	2.6
1976	−14.9	499	15.3	0.4
1977	−21.3	553	18.6	−2.7
1978	−17.5	589	21.5	4.0
1979	−16.6	1125	43.5	26.9
1980	−26.3	1955	83.4	57.1
1981	−28.4	2272	108.4	80.0
1982	−36.9	2136	113.7	76.8
1983	−26.4	2142	122.2	95.8
1984	−27.9	1789	108.4	80.5
1985	−30.4	1388	88.9	58.5
1986	−49.6	694	48.6	−1.0
1987	−45.3	506	37.9	−7.4
1988	−34.6	413	32.5	−2.1

Norway could have spent much more of the oil incomes than it actually did.

Comparing the last columns of Tables 7.6 and 7.9, we conclude that only in 1977 and 1987 did Norway have deficits on the current account that are similar to the 'warranted' deficit given by a constant debt-ratio target. According to this spending rule, if all the permanent income from the petroleum wealth had been spent, Norway should have had larger deficits (during the 1970s) or smaller surpluses (1980–85) than actually happened. Even after the oil price collapse when petroleum wealth was dramatically reduced, the deficit on the current account could have been larger than that which materialized. Given the large deficits on the current account that Norway had from 1975 to 1978 it would probably not have been feasible to spend much more due to lack of domestic resources. From 1982 to 1984 when unemployment was high and capacity utilization low, spending could have been higher due to the very optimistic expectations on oil prices that prevailed after OPEC II. In 1989, after having experienced macroeconomic adjustments policies to lower permanent incomes from petroleum wealth, one may, however, be pleased that spending was not higher in previous years.

Figure 7.14 shows graphically the warranted, oil-fund and permanent income in terms of the current account. The difference between the graph

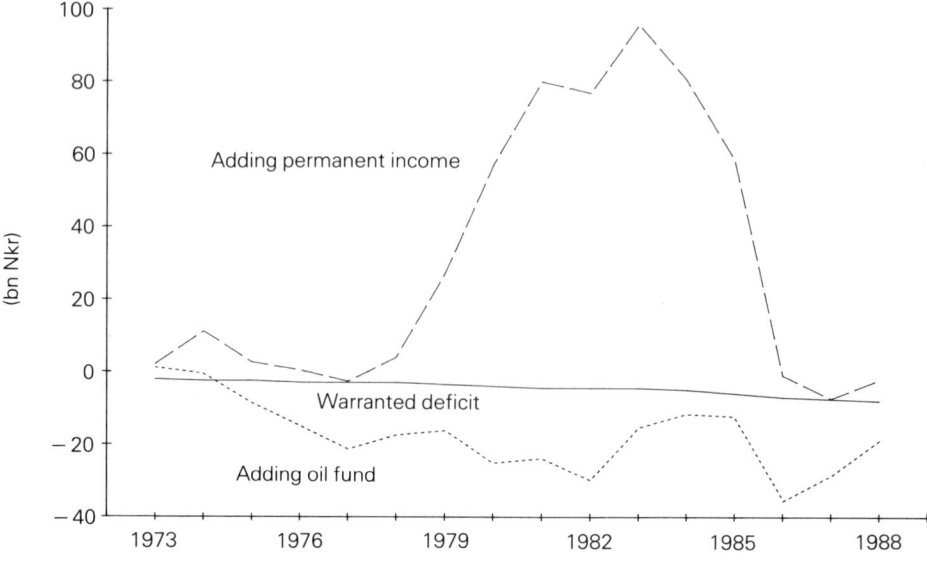

Fig. 7.14 Mainland current account.

for permanent income and warranted deficit shows how much more Norway could have spent, while the difference between the warranted and oil-fund path shows the annual amount of excess spending in terms of current account deficits. Actual spending has been somewhere between the two extremes, but closer to the oil-fund scheme than spending according to permanent income would have allowed.

7.6 CONCLUSIONS

This chapter focuses on the macroeconomic consequences of spending the rent from petroleum production. We started by reviewing some macroeconomic indicators for selected European countries among which some rely on extraction of exhaustible resources. With regard to the Norwegian experience, it is fair to say that the structural change from tradables to non-tradables did occur according to the theory of Dutch disease. However, the effect was moderate due to the fact that total supply was affected by that structural change. It also turns out that among the countries that received a resource rent (the UK, the Netherlands and Norway) macroeconomic policies differed substantially from the mid-1970s and onwards. Thus, policy objectives seem to have an important role to play. However, the developments after the 1986 oil price collapse weakens this argument, as Norway during 1989 reached unemployment rates close to the OECD average for the first time since the early 1970s. This fact was indirectly explained in section 7.4 where we illustrated the severe adjustments problems Norway would have faced without oil. The low oil prices after 1985 almost eliminated the resource rent.

Due to price uncertainty and the fact that oil and gas are non-renewable resources, much attention has been given to the problem of how to manage these resources optimally. In section 7.5 we presented two schemes that have been discussed in Norway. The first scheme, where spending is restricted to the real return of accumulated oil, is based on extreme risk aversion. If you regard non-extracted oil as part of your wealth, this scheme says that your spending should ignore that part of your total wealth completely. Only the rent that has accrued due to extraction should be taken into account. If this policy had been followed, Norway would have experienced economic development very much like the OECD area from 1975 to the early 1980, but would have done better in terms of foreign debt and employment compared to the actual figures, particularly in the later 1980s. The other scheme is based on a permanent income approach. The real rate of return on the value of all future oil rents ('oil wealth') can be spent without affecting the wealth itself. According to this scheme, if a non-renewable resource is discovered you would start consuming immediately thus relying on loans in an early period. If this policy had been followed, Norway should have spent even more than she did after 1975. This would have resulted in much higher foreign debt than was actually experienced. However, 'the perfect capital market' assumption of the permanent income life-cycle theory is probably not valid for this period, as some kind of credit rationing prevailed in international financial markets. As actual spending has been somewhere between these two schemes a preliminary conclusion may be that Norwegian spending policies have been a compromise between a scheme based on extreme risk aversion and the permanent income case.

ACKNOWLEDGEMENTS

Comments from the editors and early discussion with Steinar Strøm and Knut Anton Mork are highly appreciated.

REFERENCES

Aslaksen, I., Brekke, K. A., Johnson, T. A., and Aaheim, A., (1990) Petroleum resources and the management of national wealth, Chapter 6 in this volume.
Byatt, I., Gjelsvik, E., Goudswaard, K. and van de Kar, H. (1988) *Energy Exploitation and Public Finances in European Countries*, Leiden University, Netherlands.

Cappelen, Å., Offerdal, E. and Strøm, S. (1985) Oil revenues and the Norwegian economy in the seventies, *Macroeconomic Prospects for a Small Oil Exporting Country* (eds O. Bjerkholt and E. Offerdal), Martinus Nijhoff, Dordrecht, Netherlands.

Corden, W. M. (1985) Booming sector and Dutch disease economics: survey and consolidation, *Protection, Growth and Trade* (ed. W. M. Corden), Oxford, UK.

NOU (1983: 27) *Petroleumsvirksomhetens framtid* (The Future of the Petroleum Activities), Ministry of Oil and Energy, Oslo.

NOU (1988: 21) *Norsk økonomi i forandring* (The Norwegian Economy in Transition), Ministry of Finance, Oslo.

Steigum, E. (1989) *Bruk av inntektene fra oljevirksomhet* (Using the Incomes from the Oil Activities), Aftenposten, 12 April, Oslo.

Wijnbergen, S. van (1985) Oil discoveries, intertemporal adjustment and public policy, in *Macroeconomic Prospects for a Small Oil Exporting Country* (eds O. Bjerkholt and E. Offerdal), Martinus Nijhoff, Dordrecht, Netherlands.

8

The resource rent for Norwegian natural gas

ROLF GOLOMBEK AND MICHAEL HOEL

8.1 INTRODUCTION

Natural gas is a non-renewable resource. This means that both for society and for a private producer there is an additional cost beyond the costs of extraction and transportation. This cost is called the Hotelling rent or the (marginal) resource rent, and reflects that increased present extraction affects the possibility and profitability of future extraction. First, this is a cost due to the limited volume of natural gas, and second because increased extraction today accelerates extraction from more expensive fields if natural gas is extracted from fields with increasing unit costs. The resource rent at time t is the value of getting one more unit of the resource at time t, evaluated at time t. The magnitude of the resource rent depends upon cumulative extraction, the rate of interest and future costs and demand.

The resource rent is tied to the normative question of optimal extraction of an exhaustible resource. Assume first that the economy is closed. Then it is well known that perfect competition gives an optimal extraction path. If, on the contrary, it is possible to export the natural resource, and the country has some market power in the foreign market, equality between domestic price and the export price for the resource (corrected for differences in costs of transportation) will not be optimal for the country (cf. the theory of optimal tariffs, e.g. Johnson, 1951).

This chapter is an analysis of optimal extraction of Norwegian natural gas when natural gas can be exported and used in Norway both as an input in energy-intensive industries and in production of gas-fired thermal power. The main purpose of the analysis is to derive a rough estimate of the resource rent for Norwegian natural gas in 1990. This estimate is part of a solution of a planning problem where the paths for extraction and exports of natural gas and the development of hydroelectric power maximize the present value of the sum of profits from exports and the internal social surplus.

8.2 OPTIMAL EXTRACTION OF NATURAL GAS

In this section we study optimal extraction of Norwegian natural gas, both over time and between different activities. This problem is discussed by Førsund and Hoel (1988) within a static framework, whereas Golombek (1989) discusses the same topics within an explicit dynamic framework.

We assume throughout the analysis that natural gas can be exported or used in Norway. Natural gas can be used domestically to produce electric power (gas-fired thermal power) and as an input in the manufacturing industry. We assume that the unit cost of extracting natural gas is independent of current extraction, but that it increases over time with cumulative extraction. Unit transportation costs are assumed to be constant, with different unit costs for transportation from the continental shelf to mainland Norway and for transportation to the export market. In addition, there is a cost of transforming natural gas into electricity. Electricity is assumed to be a non-traded good, and can be produced both by hydroelectric power and by gas-fired thermal power.[1]

Electricity production in Norway has so far consisted entirely of hydroelectric power. Although the hydroelectric power capacity may be increased somewhat in the future, introduction of gas-fired thermal power as an additional electricity source is now being discussed. In addition to this domestic use of natural gas, we assume that the energy-intensive industry may use natural gas in the future (in addition to electricity).[2] Other direct domestic use of natural gas is not likely in the foreseeable future.

Norway has some market power in the European gas market. This means that Norway faces a downward-sloping demand curve for natural gas, and this curve is taken into consideration when determining the optimal export volumes. The demand schedule abroad may reflect the end users' willingness to pay either directly, or via a regulated distribution company (Golombek and Vislie, 1985; Hoel and Vislie, 1987).

Norway faces the following planning problem: find paths for export of

---

[1]Net exports of electricity from Norway have so far been negligible. However, it is possible that electricity exports will become more important in the future. Introducing that via a downward-sloping export demand function for electricity would not change our qualitative results. Quantitatively, this change would have the same effect as a faster growth of the domestic demand for electricity, which we have analysed.
[2]In Norway, the development of the energy-intensive industry is to a large extent politically governed. The price of electricity to this industry is subsidized, and the quantity of electricity is regulated. In our analysis it is assumed that the quantity of electricity to this industry stays constant at the present level. Any additional energy use by this sector must therefore come from natural gas or oil.

natural gas (x_F), domestic use of natural gas in electricity production (x_E) and directly in energy-intensive industry (x_D), and investments in hydroelectric power (J),[3] such that the present value of the export profit and the internal social surplus is maximized. The planning period is finite, and goes from 0 to T.

Formally, we may formulate the planning problem as follows (in continuous time, with t denoting time)

Maximize

$$\int_0^T [U^E(K(t) + bx_E(t), t) + U^D(x_D(t), t) + p_F(x_F(t), t)x_F(t)$$
$$- C^X(x_E(t) + x_D(t) + x_F(t), S(t), t)$$
$$- C^N(x_E(t) + x_D(t), t) - C^F(x_F(t), t)$$
$$- C^g(bx_E(t), t) - C^h(K(t), t)]e^{-rt}dt$$

given that

$$\dot{S}(t) = -(x_E(t) + x_D(t) + x_F(t)) \qquad (8.1)$$

$$S(0) = S_0 \qquad S(T) \geq 0$$

$$x_E(t) \geq 0 \qquad x_D(t) \geq 0 \qquad x_F(t) \geq 0$$

$$\dot{K}(t) = J(t) \qquad J(t) \geq 0 \qquad K(0) = K_0 \qquad (8.2)$$

Equation 8.1 expresses that at each point in time the decrease in the stock of natural gas (S) is equal to total extraction. Moreover, the three uses of natural gas must be non-negative. Equation 8.2 shows how the hydroelectric power production (K) grows over time (it can never decrease).[4]

In the maximization problem above, U^E and U^D represent the domestic utility of electricity and direct use of natural gas respectively. The total cost functions are as follows:

C^x = total costs of natural gas extraction;
C^h = total costs of production of hydroelectric power;
C^g = total costs of transformation of natural gas to electricity;
C^N = total transportation costs to the Norwegian mainland;
C^F = total transportation costs to foreign buyers.

Finally, p_F is the export price of gas, and the parameter b converts units between natural gas and electricity ($b = 5$ kWh per m^3).

The necessary conditions for an optimal solution are given in Appendix

---

[3] Capital and investments in hydroelectric power are measured in terms of electric energy.
[4] The assumption of non-decreasing hydroelectric power production follows from the two assumptions of non-depreciating capacity and full capacity utilization. Both of these assumptions are reasonable approximations for the Norwegian hydroelectric sector.

A. Before reporting the main results, it is useful to introduce some additional notation.

Let the export revenue function at time t be

$$a(x_F(t), t) = p_F(x_F(t), t)x_F(t) \qquad (8.3)$$

Interpreting the curves for marginal willingness to pay for natural gas and electricity in Norway as demand curves, domestic equilibirum prices for natural gas (p_D) and electricity (p_E) are given by

$$p_D(t) = U^{D\prime}(t) \qquad (8.4)$$

$$p_E(t) = U^{E\prime}(t) \qquad (8.5)$$

The following conditions must hold for an interior optimal solution (cf. Appendix A):

$$p_D = C_1^{x\prime} + C^{N\prime} + \alpha\, e^{rt} \qquad (8.6)$$

Direct domestic use of natural gas should be determined such that the domestic gas price (i.e. the marginal willingness to pay) is equal to the sum of the marginal cost of extraction, the marginal cost of transportation and the term αe^{rt}. The last term, which is the Hotelling rent or the (marginal) resource rent, is the value at time t of geting one more unit of the resource (with the same marginal cost as the last unit extracted).

If, along the optimal path, cumulative production at time T is less than the initial stock of natural gas, the resource rent at time T is zero. Moreover, the present value of the resource rent (i.e. α) is fixed over time if marginal cost of extraction is independent of cumulative production. This means that the resource rent is always zero if cumulative production at time T is less than the initial stock of natural gas and the marginal cost of extraction is independent of cumulative production.

The rule for determining the volume of exported natural gas is

$$a^\prime - C_1^{x\prime} + C^{F\prime} + \alpha\, e^{rt} \qquad (8.7)$$

The exported volume of natural gas should be determined such that the marginal income from export is equal to the marginal costs of extraction and transportation and the resource rent.

It is easy to see that Equations 8.6 and 8.7 yield

$$p_D - C^{N\prime} = a^\prime - C^{F\prime} \qquad (8.8)$$

The domestic price of natural gas net of marginal cost of transportation to the Norwegian mainland should be equal to marginal income from export net of marginal cost of transportation to the market abroad. This means that the export price of natural gas should be greater than the domestic price of natural gas even if the two marginal costs of transportation has been equal. In fact, marginal transportation costs are higher to foreign buyers than to the Norwegian mainland. This strengthens the conclusion that the domestic gas price should be lower than the export price.

Turning to the optimal production of electric power, we find (for an interior solution)

$$p_E = (C_1^{x'} + C^{N'})/b + C^{g'} + \alpha\, e^{rt}/b \qquad (8.9)$$

or, using Equation 8.6,

$$p_E = p_D/b + C^{g'} \qquad (8.9')$$

All variables in Equation 8.9 are expressed in value per kWh. The price of electricity (the marginal willingness to pay for electric power) should be equal to the marginal long-run cost of gas-fired thermal power, i.e. the long-run marginal costs of extraction, transportation, transformation of natural gas into electric power and the current resource rent. This means that the price of electricity measured in m^3 should be equal to the sum of the domestic price of natural gas and the cost of transforming natural gas into electric power (cf. Equation 8.9').

Electricity can also be produced by hydroelectric power. If it is optimal to increase the hydroelectric power capacity ($J(t) > 0$), the equilibrium price of electricity should be equal to the long-run marginal cost of hydroelectric power:

$$p_E = C^{h'} \qquad (8.10)$$

From Equations 8.9 and 8.10 we see that when the electricity system is optimally expanded, the long-run marginal cost of hydroelectric power is equal to the long-run marginal cost of gas-fired thermal power.

So far, we have only considered an interior solution of the planning problem. However, the optimal solution may be a corner solution, at least for some periods. In general it may not be optimal to use all three natural gas activities from time 0. Moreover, the historically given long-run marginal cost of hydroelectric power may be higher than the long-run marginal cost of gas-fired thermal power at time 0. In this case it is not optimal to expand the hydroelectric power capacity at time 0. The price of electricity should now (only) be equal to the long-run marginal cost of gas-fired thermal power. But as the long-run marginal cost of producing natural gas increases over time, there may exist a point of time τ, $0 < \tau < T$, where it is optimal to expand the hydroelectric power capacity. Then from time τ the two long-run marginal costs should be equal.

Finally, the development in the resource rent is given by

$$\frac{d}{dt}(e^{rt}\alpha(t)) = r(e^{rt}\alpha(t)) + C_2^{x'} \qquad (8.11)$$

The path of the resource rent is ambiguous when the marginal cost of extraction depends positively on cumulative production. However, if costs of extraction are independent of cumulative production, the resource rent will increase over time. In the more general case the sum of the extraction cost and the resource rent will increase over time under reasonable assumptions, see e.g. Hoel (1978).

8.3 THE RESOURCE RENT FOR NORWEGIAN NATURAL GAS

Unlike the theoretical exposition in section 8.2, our numerical calculations were done in discrete time. The length of each period was set equal to 5 years, and the planning horizon to 16 periods, i.e. 80 years. We computed the resource rent for Norwegian natural gas with the help of GAMS, which can solve complex mathematical programming models (see Brooke, Kendrick and Meeraus (1988) for a documentation of GAMS). The program computed the paths for extraction of natural gas (distributed between the different activities) and investments in hydroelectric power that solve our problem. It is then easy to find the price for electricity, the domestic price for natural gas and the export price. Finally, the resource rent follows from the necessary conditions for the solution to the optimization problem.

As opposed to the crude oil market, there is no spot market for natural gas in Europe. An important aspect of the European natural gas market is the limited number of agents on each side of the market. Algeria, the USSR, the Netherlands and Norway are the four major producers, whereas Ruhrgas (in West Germany) and British Gas are the two major buyers. Natural gas is sold under long-term contracts which specify price and volume. The price of natural gas is usually linked to prices of competing energy products, such that the price of natural gas increases when, for example, the price of crude oil increases; see Russell (1983) and Golombek and Hoel (1987) for a general discussion of natural gas contracts.

Under bargaining, rational agents use their strategic position (bargaining power) in the market: our assumption of an ordinary strictly decreasing demand for exported natural gas is clearly a rather drastic simplification of this complex bargaining situation. The observed market solution may be the outcome of a very complex game where prices and quantities are endogenous variables. The equilibrium of the game changes each time an exogenous variable shifts. If, for example, a producer's marginal cost shifts, we get a new equilibrium, and both the export price and the exported volume of natural gas will change. It may be possible to construct a downward-sloping curve in a price-quantity diagram by shifting exogenous variables; each point on the curve is the outcome from one specific game. But this curve is not an ordinary demand curve.

In spite of the game-theoretic considerations above, we simplified the modelling of the export market by assuming an ordinary decreasing demand curve. The long-run demand elasticity was determined by applying the classical conjectural variation theory: assume that the Norwegian export volume increases by 1%. Then the more the other sellers increase their volume as a response to the initial quantity increase, the more the price of Norwegian gas decreases. In particular, the two corner cases may be zero response (the Cournot case) and a 1% increase from all other sellers (the collusion case). In our reference scenario, we have chosen an intermediate case.

The export demand function is specified as

$$x_F(t) = \frac{y_F(t)}{d_F}\left(c_F - \frac{p_F(t)}{q_F(t)}\right) \qquad (8.12)$$

where c_F and d_F are constant positive parameters, and $y_F(t)$ and $q_F(t)$ are two exogenous variables, to be explained below. This demand function may be inverted to give

$$p_F(t) = q_F(t)\left(c_F - \frac{d_F}{y_F(t)} x_F(t)\right) \qquad (8.13)$$

The parameters c_F and d_F are chosen so that a gas price of Nkr0.70[5] per m^3 gives an export volume equal to 25 billion m^3 natural gas, and so that the direct price elasticity at this point is -4.

The exogenous variables y_F and q_F take care of the effects of changes in income and oil prices on the demad for natural gas. The demand curve shifts proportionally outwards due to economic growth (increased y_F); see Equation 8.12. A higher price on crude oil (increased q_F) shifts the demand curve proportionally upwards, see Equation 8.13.

The development of y_F reflects the development of economic growth abroad multiplied by an income elasticity. In our reference scenario it is assumed that y_F has a constant annual growth rate equal to 1.6%, reflecting an annual GNP growth abroad equal to 2% and an income elasticity equal to 0.8.

The development of q_F reflects the development of the price of crude oil multiplied by the ratio between a cross-price elasticity and the direct price elasticity (measured positively).[6] This ratio gives the total effect on the gas price from a 1% increase in the price of crude oil (for a given volume of gas exports from Norway). In our reference scenario, we have assumed that q_F has a constant annual growth rate equal to 0.225%, reflecting an oil price increase equal to 1% per year and a cross-price elasticity equal to 0.225 times the direct price elasticity for gas (measured positively). In other words, if the oil price increases by 1% and the price of Norwegian gas exports at the same time increases by 0.225%, the foreign demand for Norwegian gas is assumed to stay unchanged.

---

[5] The exchange rate for Nkr (= Norwegian krone) is about Nkr6.50 per $US (in June 1990).
[6] Strictly speaking, it is the price of oil products rather than the price of crude oil which is relevant for the demand of gas. However, as long as there is a stable relationship between these two prices we use the price of crude oil, provided the cross-price elasticity is defined relative to this price.

Linear demand functions similar to Equation 8.12 are used also for direct domestic demand for natural gas and for domestic electricity demand. These demand curves also shift outwards and upwards over time depending on the development of the oil price and the domestic economic growth (2% annual GNP growth in the reference scenario).

The parameter values in the reference scenario are listed in Appendix B. Most of the values are taken from Aaheim (1987). None of the functions are estimated. Variables are measured in 1986 prices not including VAT. We refer to Golombek (1989) for a more complete discussion of the choice of the parameters in the reference scenario.

The results for the reference scenario are given in Figs 8.1–8.3. Consider first the gas quantities. From Fig. 8.1 we see that the direct use of domestic gas stays at a low level throughout the planning period. The gas used for electricity production,[7] on the other hand, grows from 3.1 billion m^3 in 1990 to 42.9 billion m^3 in 2065, i.e. an average annual growth rate equal to 3.6%.[8] The exported volume of gas increases from 15.7 billion m^3 in

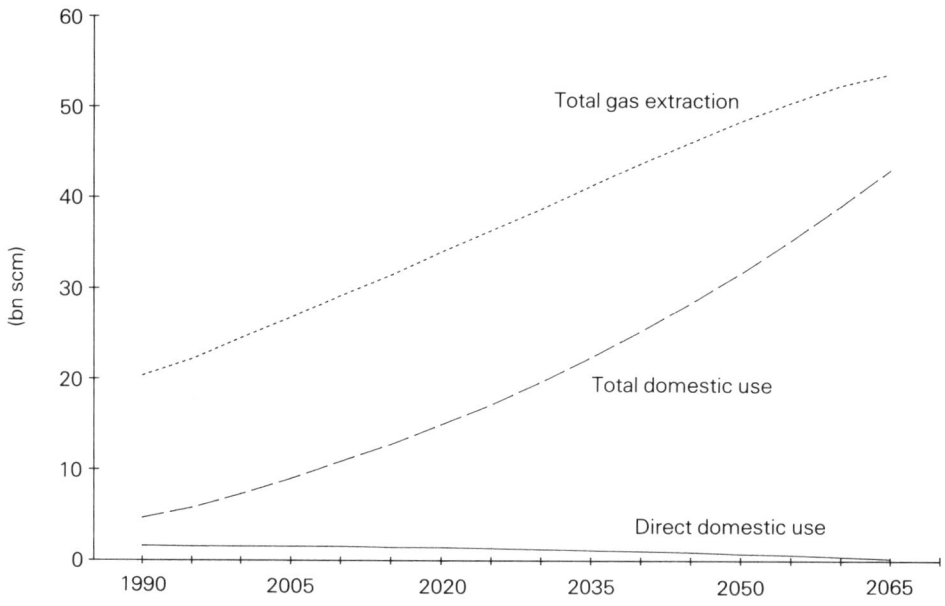

Fig. 8.1 Gas extraction and domestic use.

---

[7]The gas used for electricity production is the difference between the two lowest curves in Fig. 8.1.
[8]To simplify notation in the figures and the text, we use the first of each 5-year period to identify the period.

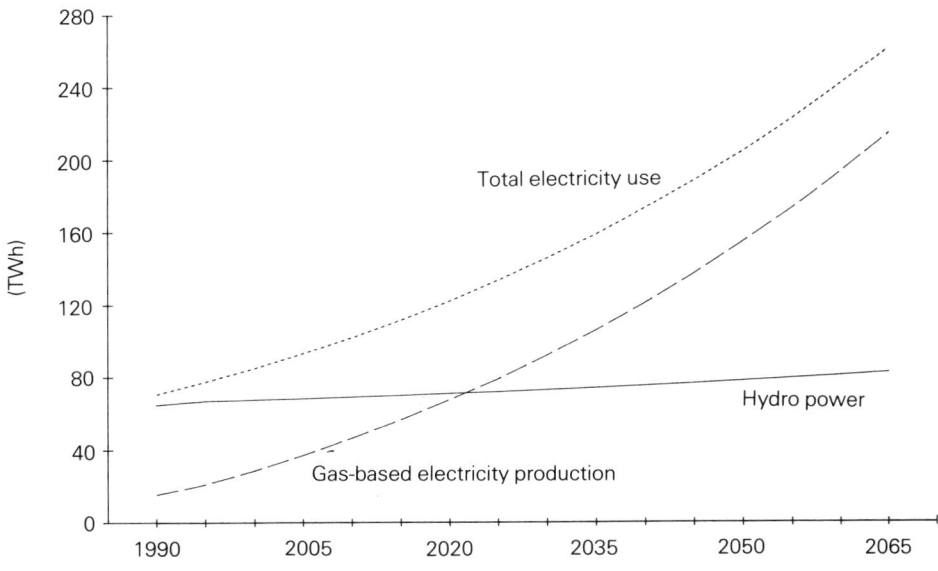

Fig. 8.2 Electricity production and consumption.

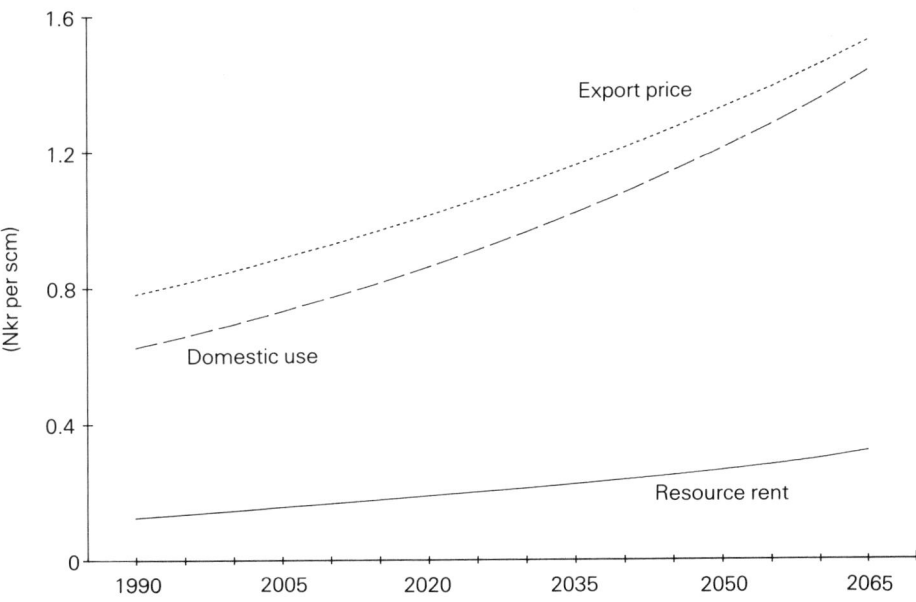

Fig. 8.3 Gas prices and resource rent.

1990 to 19 billion scm after 30 years,[9] after which gas exports decline somewhat.[10]

The rapid growth of gas used for domestic electricity production is also reflected in Fig. 8.2. The gross production of gas-fired thermal power is 15.5 TWh in 1990. This is 19% of total production of electric power. After 80 years this share has increased to 72%. Total electricity demand has an average annual growth rate equal to 1.75%, while hydroelectric power production has only a modest growth throughout the period.[11] The increase in electricity demand is slightly less than the GNP growth, in spite of an assumed income elasticity of 1 and a rising oil price. This reflects the result that the electricity price grows over time: the price of electricity for general use starts at Nkr0.29 per kWh in 1990,[12] and rises throughout the period to Nkr0.47 per kWh in 2065.

The development of the resource rent and of gas prices is given in Fig. 8.3. The resource rent increases from Nkr0.126 per m^3 (i.e. Nkr0.025 per kWh) in 1990 to Nkr0.319 in 2065. In most of the sixteen time periods, the resource rent as a share of the domestic price of natural gas in Norway is approximately 20%. The domestic price of natural gas increases from Nkr0.626 to 1.434 per m^3 during the simulation period, whereas the export price increases from Nkr0.784 to 1.523 per m^3.[13]

To test the robustness of our result, we changed the value of several parameters. (The solution in the first period did not depend much on the length of each period.) The impacts on the resource rent in the first period (which includes 1990) are summarized in Table 8.1.

In our reference scenario, the resource rent is Nkr0.126 per m^3 in 1990. A reduction in the initial stock of natural gas (3000 billion m^3) by 33% raises the resource rent in 1990 by Nkr0.007. The difference between the two scenarios increases over time, the difference is Nkr0.228 in 2070. If the initial stock of natural gas is greater than 3570 billion m^3, then the resource rent is Nkr0.125 instead of 0.126 in 1990. In the last period the resource rent is now zero, which reflects that cumulative production is less than the initial stock of natural gas. In other words, the resource rent in 1990 is

---

[9]Exported volume of natural gas from Norway was in 1988 almost 30 billion m^3, i.e. considerably higher than the optimal level according to our analysis.
[10]The volume of exported gas is given by the difference between the two upper curves in Fig. 8.1.
[11]The sum of the two lower curves in Fig. 8.2 add up to somewhat more than total electricity demand, due to an assumed 12% loss in the transmission and distribution of electricity.
[12]This price is Nkr0.04 less than the price of electricity in 1986. There is, however, a lower limit on the price of electricity: the long-run marginal cost of gas-fired thermal power net of the resource rent is Nkr0.26 per kWh in 1990.
[13]The difference between the two lower curves in Fig. 8.3 is the unit cost of extraction and transportation (to the Norwegian mainland) of gas, which increases with cumulative extraction. The difference between the two upper curves in Fig. 8.3 reflects the difference in transportation costs beween deliveries to the Norwegian mainland and exports, and also the difference between marginal revenue of exports and the export price.

Table 8.1 The resource rent and the domestic price of natural gas in 1990 (Nkr per scm)

		Resource rent	Gas price
1.	The reference scenario	0.126	0.626
2.	The initial stock of natural gas is 2000 billion m^3 (instead of 3000)	0.133	0.633
3.	The initial stock of natural gas is 3570 billion m^3 or more (instead of 3000)	0.125	0.625
4.	A halving of the slope of the function relating the marginal cost of extraction to cumulative extraction	0.093	0.593
5.	The initial extraction cost plus the cost transportation to the Norwegian mainland is Nkr0.60 per scm^3 (instead of 0.50)	0.113	0.713
6.	7% rate of interest (instead of 5%)	0.070	0.670
7.	A doubling of annual rates of income growth	0.229	0.729
8.	A halving of all annual rates of income growth	0.093	0.593
9.	The export price of 1990 is Nkr0.50 when demand is equal to 25 billion m^3 (instead of Nkr0.70)	0.074	0.574
10.	Norwegian exports can increase without any response from other producers	0.127	0.627
11.	Quadratic cost function for extraction of natural gas	0.106	0.606
12.	Environmental cost of domestic use of natural cost equal to Nkr0.50 per m^3	0.105	1.105

almost independent of the initial stock of natural gas.

To gain some more information about the relationship between the resource rent and the marginal cost of extraction, we considered a case where the marginal cost of extraction depended less on cumulative production. More precisely, we let the rise in the marginal extraction cost implied by a given cumulative extraction be half of what it was in the reference scenario. In this case the resource rent in 1990 decreased from Nkr0.126 to 0.093. Finally, the resource rent is Nkr0.113 (i.e. only slightly reduced) when both the initial marginal cost of extraction and fixed marginal cost of transportation to Norway increase with Nkr0.05.

We also studied the impact on the resource rent of different rates of interest. When we increased the rate of interest from 5% (the reference scenario) to 7%, and increased all other costs by 20%, the resource rent in 1990 decreases to Nkr0.07. This effect reflects that the increased rate of interest decreases the influence of the future. The cost of using more natural gas today, which alternatively can be used later, therefore decreases. Note, however, that the long-run marginal cost of producing natural gas in 1990 increases from Nkr0.626 to 0.67.[14]

The importance of the magnitude of the annual rates of economic growth is shown in scenarios 7 and 8. By doubling all annual rates of income growth, the level of the resource rent also doubles in the first 15 years. However, if we reduce all growth rates to half of the rates of the reference scenario, then the resource rent in 1990 decreases from Nkr0.126 to 0.093.

Scenario 9 shows the effect of a strong decrease in the price of crude oil prior to 1990: if the price of crude oil decreases so much that the export price of natural gas in 1990 decreases from Nkr0.70 to 0.50 when demand is equal to 25 billion m^3 then the resource rent in 1990 is equal to 0.074NKr. The importance of a different price elasticity for Norwegian export is, however, marginal: scenario 10 shows the impact when increased Norwegian export does not meet any response from the other sellers. The opposite case, with full response from the other producers, also has a very small impact on the resource rent in 1990, i.e. the degree of competition is of minor importance.

We also studied the importance of linear functions. First, we introduced a quadratic (convex) cost function for extraction of natural gas. The long-run marginal cost of extraction was assumed to be equal to the values of the reference scenario both when cumulative production is zero and when it is 2000 billion m^3. The resource rent is now Nkr0.106 in 1990, i.e. only slightly lower than in the reference case. In a similar way we introduced a quadratic cost function for the marginal cost of hydroelectric power production. The importance of this change was marginal.

Both a longer planning horizon and a scrap-value function, i.e. a function that measures the value of the stock at time T, will increase (or leave constant) the initial resource rent. In both cases Norway has an incentive to postpone production. Reduced production means increased price, which implies increased resource rent (since the marginal costs of extraction and transportation are independent of the produced volume in the first period). However, our results imply that the empirical importance of these two aspects is limited.

Gas-fired thermal power generates both carbon dioxide (CO_2) and nitrogen oxides (NOx). The existence of negative external effects from using natural gas decreases the resource rent (price minus marginal costs of extraction, transportation and environment). We have studied the effects of including an environmental cost of using natural gas domestically. One could argue that it is production of natural gas, and not domestic use, which is important for the environment, since for example, CO_2 emissions are indepedent of whether the gas is burnt in Norway or abroad. However, it is likely that future international regulations of, for example, CO_2

---

[14]Farzin (1984) provides a general discussion of the relationship between the rate of interest and the price of natural resources.

emissions will be on domestic emissions, and not on emissions related to goods that a country has exported. The resource rent decreases with only Nkr0.021 in 1990 when we introduce an environmental cost on domestic gas use equal to Nkr0.50 per m^3, i.e. the long-run marginal cost of natural gas increases with Nkr0.479 in 1990. As increased interest in improved environment may raise foreign demand for Norwegian natural gas, the impact from the negative externalities on the resource rent in 1990 may be quite small.

Finally, we want to find a possible floor for the resource rent in 1990. With quadratic cost function in gas extraction, a lower level of marginal cost of extraction (Nkr0.522 when cumulative production is 2000 billion m^3), 7% rate of interest, 50% decrease in annual rates of income growth and a 'big' initial stock of natural gas, the resource rent in 1990 is only Nkr0.022.

The main results for the resource rent are summarized in Table 8.1.

8.4 CONCLUDING REMARKS

We have shown that optimal extraction and use of natural gas must satisfy the following conditions:

1. Total domestic use of natural gas should be determined such that marginal willingness to pay is equal to the sum of the marginal cost of extraction, the marginal cost of transportation and the resource rent.
2. Exported volume of natural gas should be determined such that the marginal revenue from export is equal to the sum of the marginal cost of extraction, the marginal cost of transportation and the resource rent.
3. In an optimal electricity system, the long-run marginal cost of hydroelectric power and gas-fired thermal power are equal. The long-run marginal cost for gas-fired thermal power is by definition equal to the sum of marginal cost of extraction, marginal cost of transportation, the marginal cost of transforming natural gas into electricity and the resource rent.

The resource rent may be interpreted as a cost which comes in addition to ordinary cost components. It reflects that increased present extraction has a detrimental effect upon the possibility and profitability of future extraction. The resource rent is caused by the limited total volume of natural gas, and by the fact that increased extraction today accelerates extraction from more expensive fields if natural gas is extracted from fields with increasing unit costs.

The main goal of our analysis of optimal extraction of natural gas and optimal expansion of the hydro-power capacity was to find an estimate of the Norwegian resource rent in 1990. We have argued that the resource rent in 1990 is significantly greater than zero: in our reference scenario the resource rent is Nkr0.126 per m^3, i.e. 20% of the estimated domestic gas price. By changing the parameters in the model, we discovered that the existence of a positive resource rent is mainly due to a limited volume of natural gas from cheap fields. The rate of interest, the growth rates in the demand functions, the initial level of the price of crude oil and the relationship between marginal cost of extraction and cumulative production are all factors of great importance for the magnitude of the resource rent. Our calculations indicate that Nkr0.02 is a possible lower level for the resource rent in 1990.

APPENDIX A

In the following we assume that the two utility functions and the export revenue function (given by Equation 8.3 in section 8.2) are concave, and that all cost functions are convex. Letting $\alpha(\cdot)$ and $\beta(\cdot)$ be our two adjoint functions, the Hamiltonian associated with the optimization problem (*) in section 8.2 is

$$H(S, K, x_E, x_D, x_F, J, \alpha, \beta, t) = [U^E(K + bx_E, t) + U^D(x_D, t) + a(x_F, t)$$
$$- C^X(x_E + x_D + x_F, S, t) - C^N(x_E + x_D, t)$$
$$- C^F(x_F, t) - C^g(bx_E, t) - C^h(K, t)]e^{-rt}$$
$$+ \alpha(-x_E - x_D - x_F) + \beta J \tag{A.1}$$

Necessary conditions for an optimal solution are

$$\frac{\partial H(\cdot)}{\partial x_E(t)} = (bU^{E'}(\cdot) - C_1^{x'}(\cdot) - C^{N'}(\cdot) - b \cdot C^{g'}(\cdot))e^{-rt}$$
$$- \alpha(t) \leq 0 \quad (< 0 \Rightarrow x_E(t) = 0) \tag{A.2}$$

$$\frac{\partial H(\cdot)}{\partial x_D(t)} = (U^{D'}(\cdot) - C_1^{x'}(\cdot) - C^{N'}(\cdot))e^{-rt} - \alpha(t) \leq 0$$
$$(< 0 \Rightarrow x_D(t) = 0) \tag{A.3}$$

$$\frac{\partial H(\cdot)}{\partial x_F(t)} = (a'(\cdot) - C_1^{x'}(\cdot) - C^{F'}(\cdot))e^{-rt} - \alpha(t) \leq 0$$
$$(< 0 \Rightarrow x_F(t) = 0) \tag{A.4}$$

$$\frac{\partial H(\cdot)}{\partial J(t)} = \beta(t) \leq 0 \quad (< 0 \Rightarrow J(t) = 0) \tag{A.5}$$

$$\dot{\alpha}(t) = - \frac{\partial H(\cdot)}{\partial S(t)} = C_2^{x'}(\cdot)e^{-rt} \tag{A.6}$$

$$\alpha(T) \geq 0 \quad (S(T) > 0 \Rightarrow \alpha(T) = 0) \tag{A.7}$$

$$\dot{\beta}(t) = - \frac{\partial H(\cdot)}{\partial K(t)} = -(U^{E'}(\cdot) - C^{h'}(\cdot))e^{-rt} \tag{A.8}$$

Equations A.2–A.8 determine $x_E(\cdot)$, $x_D(\cdot)$, $x_F(\cdot)$, $J(\cdot)$, $S(\cdot)$, $\alpha(\cdot)$ and $\beta(\cdot)$. Finally, $K(\cdot)$ follows from Equation 8.2 in section 8.2 and $K(0) = K_0$. It is possible to prove that if the optimal solution has a jump in the state variables, then the jump must be at time 0.

Consider first conditions A.2–A.8 for the case of an interior solution. From Equations A.3 and 8.4 we immediately obtain Equation 8.6 in section 8.2.

To interpret the term $\alpha(t)e^{rt}$, let $W(0)$ be the maximum value of the sum of the export profit and the internal social surplus at time 0. For the current problem we then have

$$\alpha(t) = \frac{\partial W(0)}{\partial S(t)} \qquad t \in [0, T] \tag{A.9}$$

In other words, $\alpha(t)$ shows how much $W(0)$ increases when Norway at time t gets one more unit of natural gas. The term $\alpha(t)e^{rt}$, which is usually called the Hotelling rent or the (marginal) resource rent, is the value at time t of getting one more unit of the resource (with the same marginal cost as the last unit extracted).

If cumulative production at time T is less than the initial stock of natural gas, the resource rent at time T is zero (cf. Equation A.7). Moreover, the present value of the resource rent is fixed over time if marginal cost of extraction is independent of cumulative production (cf. Equation A.6). This means that the resource rent is always zero if the marginal cost of extraction is independent of cumulative production and cumulative production at time T is less than the initial stock of natural gas.

From Equation A.4 we obtain the rule for determining the volume of exported natural gas, i.e. Equation 8.7 in section 8.2.

Turning to the optimal production of electric power, Equation A.2 and Equation 8.5 from section 8.2 yields optimal production of gas-fired thermal power, i.e. Equation 8.9 in section 8.2.

Electricity can also be produced by hydroelectric power. Equation A.5 yields that $\beta(t)$ is equal to zero when it is optimal to increase the hydroelectric power capacity ($J(t) > 0$). Then, and only then, is the optimal price of electricity equal to the long-run marginal cost of hydroelectric power (cf. Equation 8.10 in section 8.2).

Finally, the development in the resource rent follows from Equation A.6, which gives Equation 8.11 in section 8.2.

APPENDIX B

The reference scenario

Demand for natural gas from the industry
    Annual rate of growth in demand due to increased price of oil   0.5%

Annual rate of growth in demand due to economic growth	0.75%

Demand for Norwegian natural gas abroad

Annual rate of growth in demand due to increased price of oil	0.75%
Annual rate of growth in demand due to economic growth	1.6%
Direct elasticity of price	−4

Demand for electricity

Annual rate of growth in demand due to increased price of oil	0.225%
Annual rate of growth in demand due to economic growth	2%
Direct elasticity of price	−0.53

Marginal cost of extraction (linear function)
Nkr0.30 per m^3 when cumulative production is 0 m^3
Nkr0.75 per m^3 when cumulative production is 2000 billion m^3

Marginal cost of transportation
To Norway Nkr0.20 per m^3
To the export market Nkr0.25 per m^3

Marginal cost of transforming natural gas into electric power
Nkr0.13 per kWh

Long-run marginal cost of hydroelectric power
Nkr0.24 per kWh in 1990
Nkr0.37 per kWh when the capacity has increased with 13 TWh

Rate of interest   5%

Loss in transmission and distribution of electricity   12%

Initial hydroelectric power capacity to households and firms   65 TWh

Initial stock of natural gas   3000 billion m^3

ACKNOWLEDGEMENTS

Comments from Geir Askeim, Morten Berg, Robin Lindsey, Øystein Olsen, Jon Vislie, David Wood and Asbjøn Aaheim are highly appreciated.

REFERENCES

Aaheim, A. (1987) Balansering av fremtidig vann- og gasskraft-utbygging (Balancing future investments in hydroelectric and gas-fired thermal power), unpublished paper, Central Bureau of Statistics, Norway.

Brooke, A., Kendrick, D. and Meeraus, A. (1988) *GAMS: a User's Guide*, The Scientific Press, Redwood, California

Farzin, Y. H. (1984) The effect of the discount rate on depletion of exhaustible resources. *Journal of Political Economy*, **92**, 841–51.

Førsund, F. and Hoel, F. (1988) Prinsipper for prising av gass til innenlandske formål (Principles for pricing of natural gas to domestic use), Memorandum from Department of Economics, University of Oslo, 3/88.

Golombek, R. (1989) *Optimal utnytting av gass – en beregning av ressursrenten for norsk gass* (Optimal use of natural gas – a calculation of the resource rent from Norwegian natural gas), Centre for Applied Research, Oslo.

Golombek, R. and Hoel, M. (1987) The relationship between the price of natural gas and crude oil: some aspects of efficient contracts, in *Natural Gas Markets and Contracts* (eds. R. Golombek, M. Hoel and J. Vislie), North-Holland, Amsterdam.

Golombek, R. and J. Vislie (1985) On bilateral Monopoly – a Nash–Wicksell approach, memorandum from Department of Economics, University of Oslo, 3/85.

Hoel, M. (1978) *Resource Extraction under some Alternative Market Structures*, Verlag Anton Hain.

Hoel, J. and J. Vislie, (1987) Bargaining, bilateral monopoly and exhaustible resources, in *Natural Gas Markets and Contracts* (eds R. Golombek, M. Hoel and J. Vislie), North-Holland, Amsterdam.

Johnson, H. G. (1951) Optimal welfare and maximum revenue tariffs. *Review of Economic Studies*, **19**, 28–35.

Russell, J. (1983) *Geopolitics of Natural Gas*, Ballinger Publishing Company, Cambridge, Massachusetts.

9

Social discount rates for Norwegian oil projects under uncertainty

DIDERIK LUND

9.1 INTRODUCTION

A petroleum-exporting nation in which a state oil company has a major role should seriously consider what discount rate, or shadow price, it should apply for future oil revenues. The present chapter addresses this problem, but does not give a definitive answer with respect to methods or numbers. Rather I shall discuss various approaches. I shall try to get the questions right, and to answer some of them. I refer to some Norwegian institutions, but I believe the discussion is useful for other nations with non-comprehensive stock markets as well.

Future oil prices are uncertain. What I am looking for is, for each future date, the value today of receiving one more unit of oil at that date. I call this the shadow price of oil (for that date), SPO for short. This should be the value for the government, i.e. how much it is willing to pay out of the treasury today in order to receive a claim on such a unit to be received in the future.

It is often useful to write the SPO as $\Phi E_0(P_t)$, where $E_0(P_t)$ is the expectation today (at time zero) of the price P at the future date t, and Φ is a discount factor. Of course, if both the SPO and the expected price are well defined, it is always possible to define Φ as their ratio. It will, however, not always be possible to write Φ as $1/(1+\rho)^t$, where ρ is a discount rate, the same for all values of t. The well-known term 'social discount rates' is used in the title, but is somewhat misleading.

Assume that each individual has a time-additive von Neumann–Morgenstern expected utility function

$$U = \sum_{t=0}^{T} \theta^t E_0 u(C_t) \qquad (9.1)$$

where E_0 is the expectation conditional on information available at time 0, $u(C_t)$ is the per-period utility of consumption in period t, and $\theta = 1/(1+d)$, where d is the rate of pure time preference and T is some time horizon, possibly infinite. Assume, moreover, that everyone has identical subjective probability distributions for all random variables, and identical information sets at all points in time. Admittedly these are restrictive assumptions, but well known from the literature.

Under the assumptions made, each individual is in principle able to tell his/her SPO. That is, based on the utility function and everything that affects its value, the individual is in principle able to calculate how much present consumption he/she is willing to give up to get a claim on oil to be received at any future date. But a precise answer depends on the specification of a large number of variables and functional forms, and the econometric problems would be enormous. Much of the literature therefore tries to circumvent the problem, and find easier methods based on market observables.

Before proceeding to specific methods, it is useful to clarify a few things that this chapter does not try to analyse. First of all I assume price-taking behaviour. This is one reason why the chapter refers to oil, not gas. Although since 1986 Norway has officially tried to support OPEC's pricing policies, I believe the assumption is reasonable in this connection.

Second, the problem of finding an SPO is related to the problem of finding an overall optimal extraction strategy for Norwegian petroleum. This other problem will not be solved, but I will comment on the relationship at the end. The issue to be discussed is the SPO under the actual extraction policy (or lack of policy), which is not the solution to any optimization problem that I might formulate.

Third, I do not consider flexible projects with option-like characteristics. Under the absence-of-arbitrage method for pricing of contingent claims, the assumption is that values of non-contingent claims are observable in the market. The SPO I look for is the value of such a non-contingent claim, and to find it is therefore in a sense a deeper problem than the valuation of contingent claims. In practice, there are some instances in which the shadow price can be used directly. But in other practical situations, the project or its alternative is flexible, and option-like aspects should be considered.

Fourth, the SPO is the value of one unit of extracted oil, with no extraction cost attached to it. This means that in a cost–benefit analysis, the SPO will be applicable to the revenue side, while separate shadow prices are applicable to the cost side, and possibly to financing. At the margin, of course, all elements of a project are additive, even under uncertainty.

The rest of the chapter is organized as follows. Section 9.2 considers current Norwegian practice, and the way it is justified. Section 9.3 considers models in which all individuals hold well-diversified portfolios of financial assets. Section 9.4 considers situations in which some or all individuals are not so well diversified. Section 9.5 considers models which resort to the idea of one representative individual. Section 9.6 summarizes. An appendix presents a formal model from which an expression for the SPO is derived.

9.2 CURRENT NORWEGIAN PRACTICE

The discounting rule in the public sector is justified in Ministry of Finance, Norwegian (1975) by a reference to Arrow and Lind (1970). The government behaves as risk neutral, i.e. it merely considers the expected values of costs and benefits, and uses the same discount rate for risky as for riskless projects. The problem with this is that the result of Arrow and Lind was explicitly derived for a situation in which project cash flows are uncorrelated with macroeconomic target variables. This is of course not true for oil projects in a nation which has a large oil sector.

It is nevertheless interesting to look more closely into current practice. The Ministry is inspired by Johansen (1967). A more recent work in the same tradition is Kartevoll, Lorentsen and Strøm (1980), KLS hereafter. Johansen's expression for the discount rate, ρ, under certainty is

$$1 + \rho = [u'(C_t)/u'(C_{t+1})]/\theta \qquad (9.2)$$

The expression 9.2 is obtained by considering small, opposite changes in C_t and C_{t+1} that have offsetting effects on a time-additive utility function. In general, $1 + \rho$ will depend on t. By assuming a constant growth rate of consumption, g, and that u' has a constant elasticity, v, this can be transformed to

$$\rho \approx d + (-v)g \qquad (9.3)$$

KLS (1980) use the estimates $d = 1\%$, $-v = 2$ and $g = 3\%$, which give $\rho = 7\%$. They also estimate the marginal productivity of capital in Norway to be close to 7%, and are pleased to find these numbers to be remarkably similar. Their conclusions are used to support the current practice of using a 7% real discount rate.

There are several problems connected with the method used by KLS, some of which are discussed by the authors. Among the more important ones are the assumption of constant returns to scale, the measurement of capital and the inadequate treatment of uncertainty. Furthermore, the authors suggest that the credit market should have made the consumer and producer rates of interest equal, without considering tax wedges. Below, I will return to the current practice in connection with a model of uncertainty.

9.3 MODELS OF MARKET DIVERSIFICATION

Just as other shadow prices for a cost–benefit analysis are found in ordinary commodity and credit markets, shadow prices for claims to uncertain cash flows are often looked for in markets for such claims, i.e. stock markets or futures and forward markets.

The theoretical justification for this practice is not straightforward. Very strong assumptions are needed for the standard welfare theory, known under full certainty, to hold under conditions of uncertainty. Basically, one needs complete markets for state-contingent claims, as laid out by Debreu (1959). However, there have been attempts to justify the use of observed required rates of return in stock markets in cost–benefit analysis by the government, as in Sandmo (1972), Schmalensee (1976) and Grinols (1985). An account in Norwegian of this literature is given by Lommerud (1984).

If a claim to uncertain cash flow is traded in a market, the standard assumption in these models is that each individual in an economy adjusts his/her holding of these claims optimally. In that case, the marginal rate of substitution between the uncertain cash flow and today's consumption is the same for everyone, and equal to the market price of the claim relative to consumption. This price is a natural candidate as the shadow price of the uncertain cash flow. If there is no traded claim to a given uncertain cash flow, one may try to find a spanning condition, i.e. express the cash flow as a linear combination of cash flows on which there exist traded claims. Then the price is equal to the same linear combination of the prices of these claims.

In practice, oil futures contracts have maturity dates no longer than a year into the future. In order to find a market value of one unit of oil to be received further into the future, one would have to consider the values of stocks of oil companies. To disentangle oil from the companies' other activities, financing, taxes and costs, and to distinguish between oil revenues in different future years, is certainly a very cumbersome task. For one thing, one needs as many different stocks as there are years with extraction. Some authors nevertheless regard this as the most promising way to go.

One approximation which seemingly simplifies the problem is to apply the single-period capital asset pricing model (CAPM) of Sharpe (1964), Lintner (1965) and Mossin (1966). The problem of estimating shadow prices for revenues in several future periods is then reduced to the estimation of a single, risk-adjusted discount rate. This simplification implies some very specific assumptions about the evolution of uncertainty over time. The other problems mentioned in the previous paragraph remain. But in addition, such a discount rate, or the related beta measure of risk, estimated on time-series data, does not provide a complete SPO, $\Phi E_0(P_t)$. The expected price in the future must be estimated separately. This is, of course, a drawback.

Several authors suggest the use of the CAPM to derive social discount rates in general: Bailey and Jensen (1972), for energy projects in general, Lind (1982), for energy importers, Anand and Nalebuff (1987), and for oil exporters, Adelman (1986). A Norwegian proponent is Christophersen (1984). There has been little discussion, however, on whether the assumptions underlying the CAPM are actually fulfilled. The next section will consider the assumption about individual and/or national diversification.

9.4 MODELS WITH POOR DIVERSIFICATION

A model describes poor diversification if one or more agents in the model have claims to uncertain income that are not traded and chosen in optimal amounts by the agents. The standard example in the literature is human capital. One cannot sell one's human capital because of the absence of enforceable contracts: slavery is forbidden, and there would also be weak incentives to work.

I will argue that oil-exporting nations face something of the same problem. If they consider selling their reserves in the ground, they will have to offer a substantial discount. The buyer perceives the political risks of subsequent nationalization, regulation and/or taxation. In Norway, the special petroleum tax introduced in 1975 is an example of such policy measures, not accounted for in any licence from earlier years. Similar taxes were introduced at the same time by the UK and other oil exporters. This has created the expectation among the oil companies that price increases will be followed by tax increases.

Another argument is offered by Adelman (1986), who writes that 'to sell part of their holdings is precisely what the exporting nations cannot do. It is politically impossible' (p. 319). Although he stresses this for non-industrialized oil-exporting nations in particular, something of the same is probably valid for Norway.

Irrespective of the reasons for poor diversification, we would like to have a shadow price to use for project evaluation in the current situation. It is an indisputable fact that most oil-exporting nations have national portfolios that are far from well diversified. If one imagines a national marginal rate of substitution between claims to future oil revenues and today's consumption, this is not likely to be equal to the relative price of such claims in international financial markets. A model which instead includes internal financial markets is considered below.

It is useful to consider an expression for the shadow price of a non-traded asset that is derived in Lund (1987). A multi-period version of the model is given in the appendix. The agent in the model can be either an individual or the government, as I shall discuss shortly. Assume for simplicity that the agent has access to riskless borrowing and lending, and chooses this optimally at an exogenous real interest rate, r. In addition the agent may have a number of tradable, optimally chosen claims to uncertain income, and a number of non-traded claims to such income. Then the value at time 0 of one unit more of an asset (tradable or non-traded) with an uncertain value P_1 to be obtained at time 1, is

$$\frac{1}{1+r}\left\{E[P_1] + \text{cov}\left[\frac{u'(C_1)}{E[u'(C_1)]}, P_1\right]\right\} \quad (9.4)$$

The expression in curly brackets is a certainty equivalent, to be discounted at the riskless rate. The risk-correction term, the covariance, is likely to be negative, since u' is a decreasing function and $\text{cov}(C_1, P_1)$ is likely to be positive. The latter is likely to be positive in part because the individual already may have claims to other units of the asset, so that C_1 is based on a budget that includes their value, and in part because most asset values empirically are positively correlated, so that other uncertain income is positively correlated with the one considered here. But it is possible to think of uncertain income that acts as insurance, and that has a positive risk-correction term.

Obviously, if the agent can choose his/her holding of the asset optimally at a given market price, P_0, at time zero, he/she will adjust it so that expression 9.4 equals P_0. Since P_0 and the probability distribution of P_1 are assumed to be exogenous, the adjustment takes place in C_1, i.e. in the composition of the portfolio which becomes the budget for C_1.

This risk measure is, of course, closely related to the beta measure of systematic risk in the CAPM. In the single-period CAPM it is wealth, and not consumption, that appears in the utility function, and only its mean and variance matter. All agents' portfolios are perfectly correlated, so that the covariance measure in expression 9.4 reduces to the covariance with the market portfolio.

A general discussion in Norwegian of the case of differently diversified agents, with particular reference to oil, is found in Bøhren and Ekern (1987).

The similar question of a shadow price for the government under poor diversification is analysed in Lund (1988), building on Sandmo (1972). In addition to the assumptions above, the model assumes that the government tries to obtain some Pareto optimum. The problem of obtaining a desired income distribution is neglected, and so is taxation. The model allows both individuals and the government to have non-traded claims to uncertain income, i.e. holdings of claims that are not optimally adjusted, in addition to traded claims. For a marginal unit of uncertain income at time 1, the shadow price at time 0 is a weighted average of the expression 9.4 for all individuals. The weights are the shares that each individual will have in the uncertain income that is received by the government and then distributed in predetermined shares. This result gives the desired theoretical justification for using covariance measures of risk in cases of poor diversification.

Consider the internal financial markets more carefully. Even though the nation as a whole is not well diversified, there may exist national financial markets with prices that reflect risks borne by the citizens. If all citizens choose their holdings of claims to future oil revenues optimally in such markets, they will all value one additional unit equally, namely equal to the market price of such a claim. This was stated in section 9.3, and the model in Lund (1988) is only a particular example of a set-up which gives this result.

Apart from the problems of disentangling the shadow price, the problem of poor diversification makes me reluctant to suggest market prices as the Norwegian government's shadow prices for future oil revenues. It is a fact that most Norwegians do not hold stock in oil companies. For any given future date, their SPOs are therefore in general different from each other, and not equal to a common market price.

The fact that not everyone holds non-zero amounts of all kinds of stock is of course not specific to Norwegian oil stocks. The same holds true for many assets in many economies. In addition, there are economies with no stock market at all, or with large sectors of the economy not traded in a stock market. All this makes it difficult to recommend stock market prices as social shadow prices.

One may interpret the zero holdings as optimal. This is done by Grossman and Shiller (1981), who write that the equation of the marginal rate of substitution to market prices for stocks 'holds for any individual consumer who has the option of investing in stocks (even if he chooses not to hold stocks)' (p. 223). I regard this as a very optimistic view of the theory's applicability. There are certainly other reasons, such as informational problems, or bounded rationality, that cause some people not to hold stocks. In addition there is the fact that some people do not have the option, because of, for example, short selling constraints or other forms of credit rationing.

In the case of zero holdings of Norwegian oil stocks there is one explanation which is attractive to a theorist. (Whether this explains much in practise, is another question.) This is that we, as citizens or residents, hold implicit claims to the government's future oil revenues. These claims cannot be traded. Since government services and taxes are not uniformly distributed across the population, the implicit claims are also likely to be unevenly distributed. Based on standard portfolio theory, it is reasonable to believe that, for any given market price of oil stocks, those who will hold positive amounts of the stocks in addition to the implicit claims are the most wealthy and the least risk-averse.

The rest will want negative additional holdings, but in Norway, short selling of stock is prohibited by law. The resulting equilibrium will have a

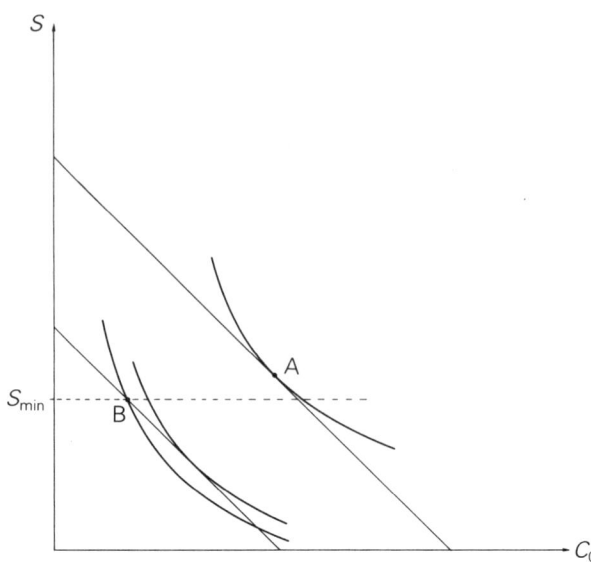

Fig. 9.1 Optimal holding of claims on future uncertain oil revenues by an individual. Case A: the individual is rich, and reaches an unconstrained optimum. Case B: the individual is poor, and reaches a constrained optimum.

higher market price of the stocks than if short selling had been allowed. A large part of the population will be at a corner solution with no oil stocks.

The situation is illustrated in Fig. 9.1, which shows two possible situations of an individual. Here C_0 is the consumption at $t = 0$, and S is claims on oil to be extracted at $t = 1$. For any portfolio the individual might hold of other assets, his/her indifference curves in the C_0, S-plane are decreasing and convex. The two decreasing lines show alternative budgets, with a given relative price of S and C_0. If this individual is wealthy, he/she attains an unconstrained optimum at A. If poor, he/she is constrained by S_{min}, the implicit claim to government oil revenues. He/she chooses a corner solution, B, at which her marginal valuation of S in terms of C_0 is lower than the market price.

If the short-selling constraint were the only deviation from standard theory, the market price of oil stocks would be an upper bound on the weighted average of all a citizen's valuations of these stocks. Equivalently, empirical estimates of the required expected rate of return on the stocks would be a lower bound on a weighted average of such rates.

The conclusion so far is that we do not have a model which justifies the use of stock market prices as sources for social SPOs. This applies both to internal market prices and to market prices in international stock markets. This does not mean, however, that the models we have cannot tell us anything. There are lessons to be learnt about qualitative aspects of the risk measure, and about current Norwegian practice as described in section 9.2.

The correct risk measure at the margin is a covariance measure. Specifically, since the social shadow price is a weighted average of expressions such as 9.4, the resulting risk measure is the covariance of the price with a weighted average of marginal utilities. One implication is that quantity risk of an oil project hardly matters in most cases. This is explained by a reinterpretation of the model discussed so far. Instead of considering one unit more of oil to be received, consider an increase in extraction which has the expected magnitude of one unit. Let P_1 in expression 9.4 be the product of the random price and the random quantity,

$$P_1 = pq \qquad (9.5)$$

where $E[q] = 1$. For most marginal project changes it is reasonable to assume that q, although risky, is stochastically independent of (the vector of) all other variables in the economy. Then

$$\text{cov}\,[u'(C_1), pq] = \text{cov}\,[u'(C_1), p]E[q] = \text{cov}\,[u'(C_1), p] \qquad (9.6)$$

The risk-correction term thus does not depend on the uncertainty in quantity. To be specific, this is valid when the term is evaluated at the margin, when C_1 is independent of q. This is similar to the standard practice in cost–benefit analysis of evaluating small changes at the shadow prices that prevail at the outset.

There are, however, some cases in which geological or technological uncertainty at the margin is correlated with C_1. This will be the case when one considers extending a large project which already makes up a significant part of the budget for C_1, and the revenues from the extension depend on the same uncertain factors as the original project.

Another lesson concerns current Norwegian practice. Equation 9.2 is based on the utility function of a single agent, perhaps the nation as a whole, and is valid under full certainty. The similar expression for a risky project derived in Lund (1987) is

$$E[R] = \frac{u'(C_0)}{E[u'(C_1)]\theta} \Big/ \left\{ 1 + \text{cov}\left[\frac{u'(C_1)}{E[u'(C_1)]}, \frac{R}{E[R]}\right] \right\} \qquad (9.7)$$

where $E[R]$ is the required expected return (one plus the required expected rate of return). It is apparent that the factor in curly brackets in most cases is less than 1, so that the required $E[R]$ exceeds the first fraction. The first fraction, on the other hand, is similar to Equation 9.2, except that the future marginal utility reappears in expected terms.

If the project had been riskless the covariance term is zero and an estimate of Equation 9.2 could be used as a required expected return. For a risky project, however, the required expected return is normally higher. The empirically estimated marginal rate of return on capital of 7% in KLS

(1980) is clearly an average of returns that includes the realizations of risk premiums. Contrary to what KLS suggest, it is not at all an optimal situation that this equals a separately obtained estimate of Equation 9.2, based on the utility function and estimates of consumption growth.

One possibility is that the separate esimate of $1 + \rho$ in Equation 9.2 at 1.07 is correct. In that case, required expected returns on risky projects should be higher than 1.07. Or the estimate of a marginal return on risky investments of 1.07 reflects the investors' required expected return on average. Then their required return on riskless investments is less than 1.07.

After 1980 credit rationing is much less important in Norway. This means that required returns on riskless investment may perhaps be observed in credit markets, not neglecting tax wedges. Since the real riskless market interest rate after tax is below 3%, it is rather unlikely that Norwegians' marginal rate of substitution for riskless projects (Equation 9.2) should be as high as 1.07. I tend to conclude that the required rate of return for riskless projects should be lower than 7%. This does not necessarily mean that I believe in 7% as an average for risky projects, but I cannot reject it. This implies that the current practice is not necessarily wrong for (some) risky projects, but it is wrong for riskless projects.

A correct practice would involve value additivity, and separate discounting for different costs and benefits since they generally have different risk characteristics. While I have focused here on the benefit side of an oil project, i.e. the revenue, a correct discounting of costs is just as important, and perhaps less discussed in the literature. The basic intuition is the same on the cost as on the revenue side, but signs are changed: it is an advantage that a factor price covaries positively with consumption and wealth. A more positive covariance implies a higher discount rate, which implies that the cost becomes an easier burden for the investor.

After making the principle clear, one will need a quantification of the cost risk. Project studies based on the CAPM sometimes suggest that costs have average riskiness, i.e. a beta of 1. Others suggest that costs are riskless. The only point to be made here is that for a Norwegian oil project it is likely that costs have a lower beta than revenues. This means that the net project value based on separate discount rates is lower than the result one gets when expected revenues and costs are discounted with any intermediate rate. The current practice implies a systematic bias in the direction of accepting too many projects. But whether this bias is counteracted by other, opposite biases, is another question.

9.5 MODELS WITH A SINGLE AGENT

The model of Lund (1988) shows that the social shadow price should be a weighted average of individual shadow prices, but it does not give an

Models with a single agent

operational method for pinning down a number. This is a difficult problem, and it may be necessary to look for simplifications. Since oil revenues in Norway mainly accrue to the Treasury, it is not too problematic to neglect distributional aspects. One can then work with models of a single agent, a representative individual.

The model should at least take into account that the national portfolio is not well diversified, and in particular that oil represents a large share. In a one-period model ($t = 0$, $t = 1$), with C_1 equal to a budget of returns on oil and n other assets, the covariance term in expression 9.4 may be approximated by

$$\mathrm{cov}\left[\frac{u'(C_1)}{E[u'(C_1)]}, p_1^{\mathrm{oil}}\right] \approx \gamma \mathrm{cov}\left[\frac{C_1}{E(C_1)}, p_1^{\mathrm{oil}}\right]$$

$$= \gamma \mathrm{cov}\left[\frac{X^0 P_1^0 + \ldots + X^{\mathrm{oil}} P_1^{\mathrm{oil}} + \ldots + X^n P_1^n}{E[C_1]}, p_1^{\mathrm{oil}}\right] \quad (9.8)$$

where $\gamma < 0$ is the coefficient of relative risk aversion. The covariance thus depends on how much oil the nation produces, multiplied by $\mathrm{var}(P_1^{\mathrm{oil}})$, and on how much the price P_1^{oil} covaries with the rest of the nation's GNP (i.e. with the rest of C_1).

The idea of approximating C_1 in this model by GNP measured as a sum of (price × quantity) for various products, has some weaknesses. These are more apparent if the lessons from intertemporal asset pricing models are taken into account. In an intertemporal, or multi-period model, one defines the optimal value function, $J_t(W_t)$, as the maximum total discounted utility as viewed from time t, as a function of the wealth. This is done in the Appendix. The risk measure becomes

$$\mathrm{cov}_0\left[\frac{J'_t(W_t)}{E_0[J'_t(W_t)]}, P_t^{\mathrm{oil}}\right] \approx \gamma \mathrm{cov}_0\left[\frac{W_t}{E_0[W_t]}, P_t^{\mathrm{oil}}\right] \quad (9.9)$$

If consumption is chosen optimally from a budget, optimal consumption can be substituted for wealth, as is done in the Appendix. However, Mankiw and Shapiro (1986) find that the covariance with wealth is a more reliable measure of systematic risk. The important point is that the first argument of the covariance should reflect the total utility of the individual, not necessarily the utility in period t. The market value of (national) wealth is likely to reflect this more adequately than per-period consumption, or per-period output value.

The most promising starting-point for an empirical investigation would be to describe the actual composition of national wealth by sector, and to estimate Equation 9.9 from data on the oil price and on the returns on assets in these sectors. In a multi-period framework one needs, among other things, an assumption on the stochastic properties of the oil price over time, and a method for valuing the oil wealth.

To conclude this section, let me comment on the relation to models that characterize an optimal extraction path of oil under uncertainty. Two such studies are Pindyck (1981) and Aslaksen and Bjerkholt (1985). One main difference between their models and the present one is the aims of the studies: my intention is to discuss the SPO under the actual extraction policy, whether optimal or not. This has led me to prefer to model extraction as exogenous.

Pindyck (1981) has a constant, exogenous discount rate for the expected value of the net revenues from extraction. This can best be explained by risk neutrality. To make it consistent with risk aversion, costs and gross revenues would have to have the same risk characteristics. But more seriously: for a nation in which oil has a large share in the national portfolio, there is no reason to believe that the risk correction in the SPO will be exogenous to the problem of the optimal extraction path. At a time when oil still has a large share, the covariance of the oil price with national wealth is high, but as the share is reduced this covariance falls. The SPO is therefore endogenous.

In Aslaksen and Bjerkholt (1985), the SPO is not explicit, but it could perhaps be derived. There are, however, other problems with their model, as I have pointed out in Lund (1986).

9.6 CONCLUSION

The Norwegian government should take risk into account when evaluating oil projects. The current practice of using a 7% discount rate applied to expected costs and revenues is not adequately justified.

It is suggested that an SPO for the government should be an average of the SPOs of the public, which are likely to be lower than the market price of oil implicit in prices in the Norwegian stock market. Quantity risk is in most cases irrelevant for the valuation of cash flows that are linear in the oil price. Costs should be discounted separately.

APPENDIX

This Appendix presents a multi-period version of the model in Lund (1987). An agent has opportunities to invest in claims with risky returns, and riskless borrowing or lending. In addition he/she has non-tradable claims to uncertain future income. The model characterizes the shadow price of obtaining one unit more of an uncertain income, i.e. one quantity unit with an uncertain value which has an exogenous probability distribution.

The time-additive utility function is

Appendix

$$U = \sum_{t=0}^{T} \theta^t E_0 u(C_t) \tag{A.1}$$

where E_0 is the expectation conditional on information available at time 0, $u(C_t)$ the per-period utility of consumption in period t, $\theta \in (0, 1)$ the utility discount factor, i.e. $\theta = 1/(1 + d)$, where d is the rate of pure time preference, and T is some time horizon, possibly infinite. The budget constraint for each time t is

$$W_t = C_t + \sum_{i=0}^{m} X_{it} P_{it} \tag{A.2}$$

where W_t is value of tradable wealth at time t, X_{it} the quantity of tradable asset i held during the period $(t, t + 1)$ and P_{it} is the unit price of this asset at t. (Dividends and similar cash payouts are neglected for simplicity, but could easily be included.) At time 0, the prices P_{0t} are known for all t, and increase exponentially, i.e. there is a known and constant interest rate, r, at which the agent is free to lend or borrow. The wealth W_t results from the holdings during the previous period, and the exogenous income at t,

$$W_t = \sum_{i=0}^{m} X_{i,t-1} P_{it} + H_t + X_{It} P_{It} \tag{A.3}$$

where H_t is some unspecified exogenous random income, while $X_{It} P_{It}$ is the exogenous random income for which I want the shadow price. That is, I want to know how much of C_0 the agent can give up if compensated by a one-unit increase in X_{It}, keeping U constant.

Stochastic dynamic programming will be used to characterize the solution. Define the optimal value function at time t as

$$J_t(W_t) = \max E_t \left[\sum_{\tau=t}^{T} \theta^{\tau-t} u(C_\tau) \right] \tag{A.4}$$

where the maximization consists in choosing a strategy. The J_t function may have other arguments than W_t, but those are not explicit here. The equation of optimality has the form

$$\begin{aligned} J_t(W_t) &= \max E_t[u(C_t) + \theta J_{t+1}(W_t + 1)] \\ &= \max_{X_{0t},\ldots,X_{mt}} \left\{ u\left(W_t - \sum_{i=0}^{m} X_{it} P_{it}\right) \right. \\ &\quad \left. + \theta E_t \left[J_{t+1}\left(\sum_{i=0}^{m} X_{it} P_{i,t+1} + H_{t+1} + X_{I,t+1} P_{I,t+1} \right) \right] \right\} \end{aligned} \tag{A.5}$$

The first-order conditions for an optimum are, for $i = 0, \ldots, m$,

$$\frac{\partial J_t}{\partial X_{it}} = u'(C_t)(-P_{it}) + \theta E_t[J'_{t+1}(W_{t+1}) P_{i,t+1}] = 0 \tag{A.6}$$

For this optimal solution, the envelope theorem gives, for all t,

$$\frac{\partial J_t}{\partial W_t} = u'(C_t) \qquad (A.7)$$

By substituting this expression, with $t+1$ instead of t, for J'_{t+1} in Equation A.6, one can derive

$$1 + r = \frac{P_{0,t+1}}{P_{0t}} = \frac{u'(C_t)}{\theta E_t[u'(C_{t+1})]} \qquad (A.8)$$

and, for $i = 1, \ldots, m$,

$$E_t\left(\frac{P_{i,t+1}}{P_{it}}\right) = 1 + r - \text{cov}_t\left[\frac{u'(C_{t+1})}{E_t[u'(C_{t+1})]}, \frac{P_{i,t+1}}{P_{it}}\right] \qquad (A.9)$$

For the non-traded claim on oil, use the envelope theorem similarly, and use the fact that for all positive s and t,

$$\partial E_t(J_{t+s})/\partial E_t(J_{t+s+1}) = \theta.$$

This gives

$$\partial J_0/\partial X_{It} = \theta^{t-1}\partial E_0(J_{t-1})/\partial X_{It} = \theta^t E_0[u'(C_t)P_{It}] \qquad (A.10)$$

Observe that for $0 \leq s \leq t$,

$$(1+r)\theta E_{t-s}[u'(C_{t+1})] = E_{t-s}[u'(C_t)] \qquad (A.11)$$

which implies that

$$(1+r)^t \theta^t E_0[u'(C_t)] = u'(C_0) \qquad (A.12)$$

Together with Equation A.10, this gives the shadow price for one unit more of X_{It}, evaluated in real terms at time 0,

$$\frac{\partial J_0}{\partial X_{It}} \bigg/ \frac{\partial J_0}{\partial W_0} = \frac{1}{(1+r)^t}\left\{E_0[P_{It}] + \text{cov}_0\left[\frac{u'(C_t)}{E_0[u'(C_t)]}, P_{It}\right]\right\} \qquad (A.13)$$

The expression 9.4 in the main text is merely a special case of this, with $t = 1$.

ACKNOWLEDGEMENT

I would like to express my thanks to Aanund Hylland for comments made.

REFERENCES

Adelman, M. A. (1986) Oil producing countries discount rates. *Resources and Energy*, 8, 309–29.
Anand, S. and Nalebuff, B. (1987) Issues in the application of cost–benefit analysis

References

to energy projects in developing countries. *Oxford Economic Papers*, 39, 190–222.
Arrow, K. J. and Lind, R. C. (1970) Uncertainty and the evaluation of public investment decisions. *American Economic Review*, 60, 364–78.
Aslaksen, I. and Bjerkholt, O. (1985) Certainty equivalence procedures in the macroeconomic planning of an oil economy, in *Macroeconomic Prospects for a Small Oil Exporting County* (eds O. Bjerkholt and E. Offerdal), Nijhoff, Dordrecht, The Netherlands, pp. 283–318.
Bailey, M. F. and Jensen, M. C. (1972) Risk and the discount rate for public investment, in *Studies in the Theory of Capital Markets* (ed. M. C. Jensen) Praeger, New York, pp. 269–93.
Bøhren, Ø. and Ekern S. (1987) Usikkerhet i oljeprosjekter. Relevante og irrelevante risikohensyn (Uncertainty in oil projects: relevant and irrelevant risk considerations). *Beta*, 1(1), 23–30.
Christophersen, Y. (1984) Samfunnsøkonomisk risiko og kalkulasjonsrente (Social risk and discount rate). *Bedriftsøkonomen*, 46, 311–13.
Debreu, G. (1959) *Theory of Value*, Wiley, New York.
Grinols, E. L. (1985) Public investment and social risk-sharing. *European Economic Review*, 29, 303–21.
Grossman, S. J. and Shiller, R. J. (1981) The determinants of the variability of stock market prices. *American Economic Review*, 71, 222–7.
Johansen, L. (1967) *Investeringskriterier fra et samfunnsøkonomisk synspunkt* (Investment criteria from a social point of view). Ministry of Finance, Oslo.
Kartevoll, T., Lorentsen, L. and Strøm, S. (1980) Kalkulasjonsrenten (The discount rate). *Sosialøkonomen*, 34(6), 9–16.
Lind, R. C. (1982) A primer on the major issues relating to the discount rate for evaluating national energy options, in *Discounting for Time and Risk in Energy Policy*, (ed. R. C. Lind), Johns Hopkins University Press, Baltimore, Maryland, pp. 21–94.
Lintner, J. (1965) The valuation of risk assets and the selection of risky investments in stock portfolios and capital budgets. *Review of Economic and Statistics*, 47, 13–37.
Lommerud, K. E. (1984) Velferdsøkonomisk perspektiv på olje og usikkerhet (Welfare economic perspective on oil and uncertainty). *Statsøkonomisk tidsskrift*, 98 (2), 41–65.
Lund, D. (1986) Some comments on the solution of a dynamic portfolio model with endogenous oil and on oil price uncertainty, Memorandum no. 14, Department of Economics, University of Oslo, Norway, 10 February.
Lund, D. (1987) *Investing in non-marketable assets*, Memorandum no. 2, Department of Economics, University of Oslo, Norway, 19 February.
Lund, D. (1988) *Social discount rates under uncertainty–a re-examination and extension of Sandmo's 'farm' model*, Memorandum no. 10, Department of Economics, University of Oslo, Norway, 6 July.
Mankiw, N. G. and Shapiro, M. D. (1986) Risk and return: consumption beta versus market beta. *Review of Economics and Statistics*, 68, 452–9.
Ministry of Finance, Norwegian (1975) *Den samfunnsøkonomiske kalkulasjonsrenten* (The social discount rate), Division of Finance, Oslo, Norway, March.
Mossin, J. (1966) Equilibrium in a capital asset market. *Econometrica*, 35, 768–83.
Pindyck, R. S. (1981) The optimal production of an exhaustible resource when price is exogenous and stochastic. *Scandinavian Journal of Economics*, 83, 277–88.
Sandmo, A. (1972) Discount rates for public investment under uncertainty. *International Economic Review*, 13, 287–302.

Schmalensee, R. (1976) Public investment criteria, insurance markets, and income taxes. *Journal of Public Economics*, **6**, 425–45.

Sharpe, W. F. (1964) Capital asset prices: a theory of market equilibrium under conditions of risk. *Journal of Finance*, **19**, 425–42.

10

The choice between hydro and thermal power generation under uncertainty

T. Ø. KOBILA*

10.1 INTRODUCTION

Norway has abundant energy supplies consisting of hydro power, crude oil and natural gas and is one of the few countries of the world in which the electricity supply is more than 99% based on hydro power. This reflects that hydro power until recently has been the cheapest source for covering a steadily increasing demand for electricity. The fall in price of crude oil in 1986 has also brought down the price of natural gas in Western Europe. This has actualized the issue of whether further expansion of the electricity supply system in Norway should be based on thermal power generation or on still unexploited hydro sources.

This is a question of marginal cost comparison which is basically very simple, but becomes less transparent when irreversibility and uncertainty aspects are taken into consideration. The effect of uncertainty on the choice between hydro and thermal power was first discussed in the classical text by Massé (1962). However, he gave conditions for the optimality of parallel development rather than the phasing in of an irreversible investment. The importance of uncertainty is exacerbated by the difference in cost structure between the hydro-power plant and the gas-fired plant. In our presentation we stylize this difference by representing the cost of hydro power solely as everlasting capital, while only the fuel cost of thermal power is taken into consideration.

We consider two sources of uncertainty. Uncertainty of future demand

---

*This is an acronym for: Iulie Aslaksen, Olav Bjerkholt, Kjell Arne Brekke, Tom Lindstrøm and Bernt Øksendal.

stems mainly from cyclical factors and temperature variations. The uncertainty of the opportunity cost of natural gas stems from the export market which is the alternative outlet for the domestic supplies of natural gas. Another important source of uncertainty, which we have not tried to incorporate in the formal analysis here, is the variations in annual output from a hydro-power system due to variations in rainfall.

The energy sector in Norway is dominated by government-owned companies. The planning of future energy supply is clearly a task for the government, although the same reasoning would apply to a large utility company. Our main motivation for approaching this problem is that traditional planning methods tend to underplay considerably the role of uncertainty and be biased towards over-investment in hydro-power capacity. This bias is more costly when other energy sources become more competitive, future demand more uncertain and new hydro power more expensive.

The methodological approach of this chapter is to represent the uncertainty as stochastic processes of Brownian motions. Such methods have in recent years received widespread attention in economic applications as a powerful tool of analysis. Although the mathematical requirements for solving these stochastic control problems are quite demanding, the character of the solutions are surprisingly simple and open to direct interpretations that could be incorporated within traditional planning procedures as rules of thumb. Realistic solutions may, however, require computer-intensive numerical processing.

We also emphasize the irreversibility aspect of the investment decision. The importance of irreversibility in investment problems under uncertainty is clearly recognized in numerous contributions in environmental economics. Arrow and Fisher (1974) conclude that when investments are irreversible, optimal investments under uncertainty should be lower as compared to a situation with reversible investment decisions. In a two-period model Fisher and Hanemann (1982) analyse the combined effect of uncertainty and irreversibility as a rationale for deferred investment. In a recent article Pindyck (1988) has derived a criterion for irreversible investments under uncertainty. In our analysis we find an explicit expression for the irreversibility premium, that is, the reduction in hydro-power investment due to uncertainty in future demand.

We address the investment timing problem in the framework of optimal stopping. This powerful tool of analysis (Øksendal, 1985; Shiryayev, 1978) is only beginning to find applications in economics. Recent applications include McDonald and Siegel (1986), Roberts and Weitzman (1981) and two recent papers on resource extraction by Olsen and Stensland (1987, 1988). Option pricing methods and optimal stopping rules, applied to petroleum projects, are found in Bjerkholt and Brekke (1988), Ekern (1988) and Paddock, Siegel and Smith (1988). Our analysis of the investment timing problem is similar to the stochastic control problems of Majd and

Pindyck (1985) and Brennan and Schwartz (1985), who derive the option value arising from the option to postpone investment.

10.2 UNCERTAINTY AND IRREVERSIBILITY: OPTION VALUE IN THE CHOICE BETWEEN THERMAL AND HYDRO POWER

We first pose the problem of whether to choose thermal power or hydro power to cover a given expansion of demand when the price of thermal energy is uncertain. A crucial assumption is that hydro-power investment is irreversible; once the investment is in place it will never pay to switch to thermal power, however low the price of the thermal energy source may fall. The decision to defer investment, however, is reversible. The formulation of this investment problem with its features of uncertainty and irreversibility can be cast in the form of a so-called **option value** problem. The essence of the option value problem in this chapter is as follows.

The owner of the hydro source has the option of producing power at a known cost by investing the required amount of capital, K. The alternative source of power is from gas-fired plants at a fuel cost of, say, q per unit of energy generated. If the fuel price is constant and known with certainty, the option is worth nothing if $q \leq r_F K$, and no investment should be made (r_F is the rate of return of a risk-free asset, and $r_F K$ has the interpretation as long-term marginal cost in the expansion of the hydro-power system). At a higher price the investment should be made and it is simple to work out that the option is worth $q/r_F - K$. If, on the other hand, the price is a stochastic process, a low price today does not rule out a higher price tomorrow. Hence, at a low price the option has a positive value, although no investment should be made immediately. Even at a price slightly higher than $r_F K$ the investment should still be withheld. At a sufficiently high price, however, the investment should be made. The derivation of the option value under uncertainty will be shown below.

Assume thus that an additional unit of demand of electricity can be covered either by an increase of hydro-power capacity at capital cost K or from gas-fired plants at a fuel cost of Q_t per unit of energy generated. The gas price is stochastic; the realization of the stochastic process Q_t at the initial time is denoted by q. Consider the cost minimization problem of covering the additional demand. The minimum expected cost is a function of time through discounting and the initial gas price q at t.

$$C(t, q) = \min_\tau E^{t,q} \left\{ \int_t^\tau Q_s e^{-r_Q s} ds + K e^{-r_F \tau} \right\} \qquad (10.1)$$

Total cost here consists of discounted fuel costs in the period from t to τ – the point in time of switching from thermal to hydro power – and the discounted capital cost at τ. The time τ is a stochastic time depending on the unveiling of the stochastic process Q_t. This means that the investment decision can be postponed until we observe the required value of Q_t. (In most precise terminology τ is restricted to the set of stopping times with respect to the process Q_t.)

Note that in Equation 10.1 fuel cost and capital cost are discounted with different discount rates. Capital cost is non-stochastic and is discounted at the risk-free rate of return r_F. The stochastic fuel cost is discounted at r_Q, the required rate of return on an asset with the same stochastic properties as the gas price Q_t. We shall comment further upon these discount rates and the appropriate rate to apply to the option value itself.

The assumptions made about the stochastic process Q_t are fairly standard, they imply that for given Q_t the probability distribution of Q_s ($s > t$) is lognormal. More precisely, we assume that the stochastic process for Q_t is a geometric Brownian motion given by

$$dQ_t = \mu Q_t dt + \sigma Q_t dB_{1t} \qquad (10.2)$$

In Equation 10.2 the first term expresses the exponential drift of Q_t while the second term represents the uncertainty of Q_t as a Brownian motion (Wiener process) B_{1t}. Equation 10.2 implies that

$$E\{Q_s|Q_t\} = Q_t\, e^{\mu(s-t)} \quad \text{and} \quad \mathrm{var}\left(\ln\left(\frac{Q_s}{Q_t}\right)\right) = \sigma^2(s-t) \quad \text{for } s > t$$

Under full certainty the cost minimization problem corresponding to (10.1) is solved by investing in hydro power as soon as the current fuel cost exceeds $r_F K$. Under uncertainty the investment criterion has a similar form, given in terms of a **reservation price**, for which we shall see that we can find an explicit expression. First, we rewrite the discounted fuel cost term of Equation 10.1 in the following way:

$$\begin{aligned}
E^{t,q}\left\{\int_t^\tau Q_s e^{-r_Q s} ds\right\} &= E^{t,q}\left\{\int_t^\infty Q_s e^{-r_Q s} ds\right\} - E^{t,q}\left\{\int_\tau^\infty Q_s e^{-r_Q s} ds\right\} \\
&= \int_t^\infty E^{t,q}(Q_s) e^{-r_Q s} ds - E^{t,q}\left\{\int_\tau^\infty E^{\tau,Q^\tau}(Q_s) e^{-r_Q s} ds\right\} \\
&= \frac{1}{r_Q - \mu}[qe^{-r_Q t} - E^{t,q}\{Q_\tau e^{-r_Q \tau}\}] \qquad (10.3)
\end{aligned}$$

Equation 10.3 simply says that the expected cost of buying gas in the period t to τ is equivalent to the expected net return from buying a right to eternal gas delivery at t and selling this right at τ. Whether or not it is optimal to sell the right to gas delivery at τ depends on the value Q_τ compared to K. Using Equation 10.3 in 10.1 gives us the following expression for minimum expected cost:

$$C(t, q) = \frac{1}{r_Q - \mu}[q\, e^{-r_Q t} - \max_\tau E^{t,q}\{(Q_\tau e^{-r_Q t} - K(r_Q - \mu))e^{-r_F \tau}\}] \qquad (10.4)$$

The maximization expression in Equation 10.4 is formally equivalent to another option value problem, i.e. the option to buy an asset with the same stochastic properties as Q_t at a price $K(r_Q - \mu)$. By reference to McDonald and Siegel (1986) we shall draw on some results from the capital asset pricing model (CAPM) of financial economics. A general assumption here is that investment options are owned by well-diversified investors who need only be compensated for the systematic component of risk. We have earlier introduced the risk-free rate of return, r_F, and the required rate of return, r_Q. (According to CAPM, the difference $r_Q - r_F$ is proportional to the correlation between the rate of return on an asset with the same stochastic properties as Q_t, and that on the market portfolio. If the correlation with the market portfolio is constant, r_Q is a linear function of σ, the standard deviation of Q_t.) As shown in McDonald and Siegel (1986), there is a discount rate, r, appropriate for each option value problem. This discount rate is given by

$$r = \gamma r_Q + (1 - \gamma) r_F,$$

where $\gamma > 1$. Hence, when the stochastic cost component has a required rate of return r_Q, and the deterministic cost component has a required rate of return r_F, the associated option value has a required rate of return r. Then Equation 10.4 can be rewritten as

$$C(t, q) = \frac{1}{r_Q - \mu} [q e^{-r_Q \tau} - \min_\tau E^{t,q}\{(Q_\tau - K(r_Q - \mu))e^{-r\tau}\}] \quad (10.5)$$

The solution to Equation 10.5 follows immediately from the optimal stopping theorem given in the Appendix. The theorem says that a maximization problem of the form as in Equation 10.5 has a solution in terms of a reservation price, i.e. the sale of the right to gas deliveries should be postponed until the gas price for the first time reaches the reservation price q^*. From the theorem it follows that

$$q^* = r^* K = \frac{\gamma(r_Q - \mu)}{\gamma - 1} K > r_F K \quad (10.6)$$

The coefficient γ, which is the same as the one that enters the expression for r given above, depends on the drift and the variance of the stochastic process Q_t. The relationships are set out in the Appendix, where it is also shown that the reservation price q^* is higher than the reservation price $r_F K$ in the case of no uncertainty.

Thus, the hydro-power project has an option value and the investment should be postponed until Q_t reaches q^*. As long as $Q_t < q^*$ there is an expected cost reduction from holding the option as it will be exercised when the gas price increases above q^*. Define $\widetilde{Q}_t$ as

$$\widetilde{Q}_t = \begin{cases} r_F K & \text{if } Q_t > q^* \\ Q_t & \text{otherwise} \end{cases} \quad (10.7)$$

The choice between hydro and thermal power generation

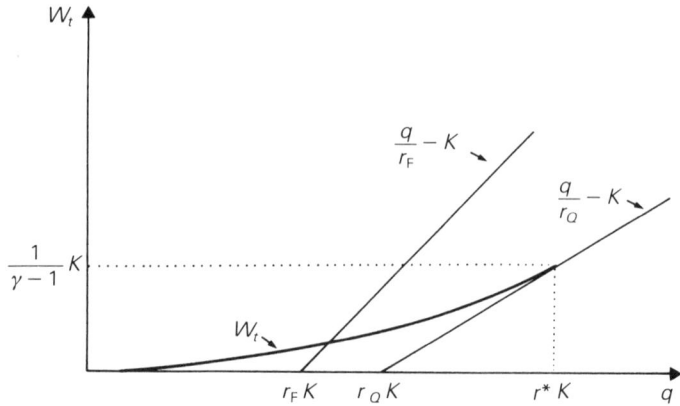

Fig. 10.1 Option value of hydro power in the case of zero drift in the gas price ($\mu = 0$).

It follows that $E\tilde{Q}_t \leq EQ_t$ since the option to switch to hydro power represents a truncation of the probability distribution of Q_t. Hence, the expected cost is reduced by holding the option.

Figure 10.1 shows the option value under uncertainty, denoted by W_t, and under full certainty, denoted by $q/r_F - K$, and illustrates two different effects of uncertainty for the case of $\mu = 0$. Under full certainty, the hydro-power investment depends on whether the constant gas price q is less than or greater than $r_F K$. In the first case the option value is zero, in the latter case it is given by the straight line $q/r_F - K$. Under uncertainty we need to distinguish the effect of uncertainty on the required rate of return, and the irreversibility effect. If Q_t has a positive correlation with the market portfolio, the required rate of return, r_Q, is higher than r_F. In this case the line $q/r_F - K$, the option value under certainty, is steeper than the corresponding line for the required rate of return r_Q. Under uncertainty the relevant comparison is between the curve W_t and the line $q/r_Q - K$. Investment in hydro power is undertaken when the curve meets the line, at $q = q^* = r^* K$. W_t is the gain from having the option of switching to hydro power at τ, as compared to a pure thermal-based expansion of the power supply, and it is formally represented by the maximization expression in Equation 10.5. From the solution given in the Appendix we have that, for $t = 0$,

$$W_t(q) = \frac{1}{\gamma - 1} K \left(\frac{q}{q^*}\right)^{\gamma} \tag{10.8}$$

For $q = q^*$ the option value is given by

$$W_t(q^*) = \frac{1}{1 - \gamma} K$$

If there is no correlation between Q_t and the market portfolio, $r_Q = r_F$, and the two straight lines coincide. In this case the effect of uncertainty is only expressed as the irreversibility effect, $r^* > r_F$, whereas the case of positive correlation gives the combined effect via the required rate of return and irreversibility. We can find an explicit expression for the irreversibility premium, that is, the reduction in hydro-power investment due to uncertainty in future demand. The irreversibility premium is given by

$$r^* - r_Q = \frac{r_Q}{\gamma - 1}$$

Under full certainty the irreversibility premium approaches zero since $\lim_{\sigma \to 0} \gamma = \infty$. The interpretation is that the gain from deferring investment vanishes when the price remains constant. When the degree of uncertainty increases, the irreversibility premium approaches infinity since $\lim_{\sigma \to \infty} \gamma = 1$.

A standard result in option value theory is that the option value is higher under uncertainty as compared to full certainty. Taking the combined effect of irreversibility and positive correlation into account, this conclusion is modified. From Fig. 10.1 we see that for $q < r_F K$, the general argument applies and the option value is higher under uncertainty. For $q > r^* K$, the value of the hydro-power project is higher under certainty, since the line $q/r_F - K$ is steeper than $q/r_Q - K$. To see this, note that the positive correlation implies that the required rate of return is higher under uncertainty, hence reducing the present value of future fuel cost. Thus the option to switch to hydro power is less valuable, and the potential gain from uncertainty is diminished. Hence, there is a q, with $r_F K < q < r^* K$, such that the potential gain from uncertainty (the difference in option value) changes from positive to negative.

If the correlation between the asset with the same stochastic properties as Q_t and the market portfolio is negative, we have that $r_Q < r_F$. In terms of Fig. 10.1 this means that the option value under uncertainty is unambiguously higher. This is intuitively reasonable since with negative correlation the market portfolio provides a hedge against gas price uncertainty.

Finally, we will clarify why the reasonable assumption of this model is that the risk-adjusted discount rate is higher than the risk-free discount rate. The profit from thermal electricity generation will be negatively correlated with export revenues, for if oil and gas prices are high, export revenues are high, and the cost of thermal power will be high, lowering the profit from power generation. If oil and gas prices are low, thermal electricity costs will be low and profits that much higher. Electricity sales, prices and profits are likely to move in sympathy with oil export revenues, which will positively influence general economic activity, and the electricity sector thus hedges its profits by buying inputs whose price covaries with output prices or revenue. Consequently, investing in hydro rather than expanding with thermal power means that profit will now be more highly correlated with oil export revenues, and that makes the electricity sector more susceptible to uncertainty, requiring a higher discount rate than the risk-free rate. Recall that the decision to invest in hydro is equivalent to selling the gas contract, which in contrast to buying gas, increases the correlation.

As a numerical illustration of the investment criterion given by Equation 10.6 let us set $r_F = 0.05$, $r_Q = 0.06$, $\mu = 0.03$ and $\sigma = 0.12$. The reservation price under uncertainty is then $0.06324K$ as compared to $0.05K$ under certainty. In other words, the hydro investment should be put off until the fuel price exceeds the long-term marginal cost by 26%.

10.3 OPTIMIZATION OF ENERGY SUPPLY OVER TIME UNDER UNCERTAIN DEMAND

In section 10.2 we considered how to cover an additional unit of demand for energy, by irreversible investment at known cost, or relying on available fuel supplies at stochastic cost. We will now consider the planning problem of the power authority when the demand of electricity is stochastic. Given that the investment is irreversible, when should the remaining reserve of hydro power be phased in?

We assume that the electricity market is in equilibrium. As before, we make very stylized assumptions about the cost structure of both hydro and thermal power. Furthermore, we ignore adjustment costs and assume that hydro-power capacity can be increased in arbitrary small amounts. We represent the demand for electricity as a stochastic process with positive drift and non-negligible variance. The demand curve is chosen simple enough to allow an explicit solution:

$$D_t = P_t^{-\varepsilon} \Theta_t \qquad \varepsilon > 0 \qquad (10.9)$$

where D_t is the demand of electricity, P_t is the price of electricity and Θ_t is a stochastic demand shift parameter given as geometric Brownian motion:

$$d\Theta_t = \alpha \Theta_t dt + \beta \Theta_t dB_{2t} \qquad (10.10)$$

Here Θ_t may be interpreted as the income effect and other factors influencing demand.

The hydro-power capacity is denoted by K_t, not to be confused with the capital cost of section 10.2. More capacity is available at increasing costs. The unit cost of another unit of capacity is given by

$$C(K_t), \; C'(K_t) > 0 \qquad C''(K_t) > 0.$$

Our control variable is additional hydro-power investment $\kappa_t \geq 0$.

$$dK_t = \kappa_t dt \qquad (10.11)$$

The alternative source of energy is thermal power (natural gas), for which we now assume a constant price q. Completed hydro-power investments are sunk cost, and maximum benefit requires the full capacity to be used. The planning problem consists in maximizing the expected discounted value of consumer surplus less costs. The consumer surplus is defined as the area under the demand curve (Equation 10.9). It is given by

$$\int_a^{D_s} \left(\frac{\Theta_s}{x}\right)^{1/\varepsilon} dx$$

In the definition of the consumer surplus the constant $a > 0$ is introduced to ensure convergence of the integral also in the case of $\varepsilon < 1$. The objective function is given by

$$H(t, \theta, k; q) = \max_K E^{t,\theta,k} \left\{ \int_t^\infty \left[\int_a^{D_s} \left(\frac{\Theta_s}{x}\right)^{1/\varepsilon} dx - C(K_s)\kappa_s \right. \right.$$
$$\left. \left. - q(D_s - K_s) \right] e^{-rs} ds \right\} \tag{10.12}$$

The values of the state variables Θ_t and K_t at time t are denoted by θ and k. The optimal value function H depends on time as well as the two state variables. The discount rate r is not necessarily related to the discount rate of section 10.2. The gas price, q, enters as a parameter. If demand at price q is higher than the hydro-power capacity, the excess demand is covered by thermal power. The equilibrium price will always ensure that the hydro-power capacity is utilized.

$$P_t = \min \left\{ \left[\frac{\Theta_t}{K_t}\right]^{1/\varepsilon}, q \right\} \tag{10.13}$$

Equations 10.10–10.12 give a problem of stochastic control. We will begin by outlining the structure of the solution, and then proceed to the mathematical formulation.

In contrast to the model of section 10.2, where the optimal strategy is to invest when the random gas price exceeds a reservation price, the optimal strategy in this model depends on two state variables, the random parameter, Θ_t, and the capacity already invested, K_t. With two state variables, the optimal investment strategy cannot be expressed in terms of a single reservation price. In two dimensions, the idea of a reservation price is represented by a curve in the (θ, k)-plane, denoted by $k = k^*(\theta)$ (Fig. 10.2).

It turns out that the stochastic control problem explained below has a singular solution. This means that the optimal investment is either zero or infinity, depending on the current realization of the process (Θ_t, K_t) in relation to the curve $k = k^*(\theta)$. The area below the curve $k = k^*(\theta)$, denoted by $\mathcal{A}$, is characterized by an infinite investment rate. This means that if the process initially starts in $\mathcal{A}$, it will immediately jump to the curve $k = k^*(\theta)$. Then the capacity is optimal for the given level of θ, denoted by θ^*. The relation $\theta^* = \theta^*(k)$ is the inverse of $k = k^*(\theta)$, and it is interpreted

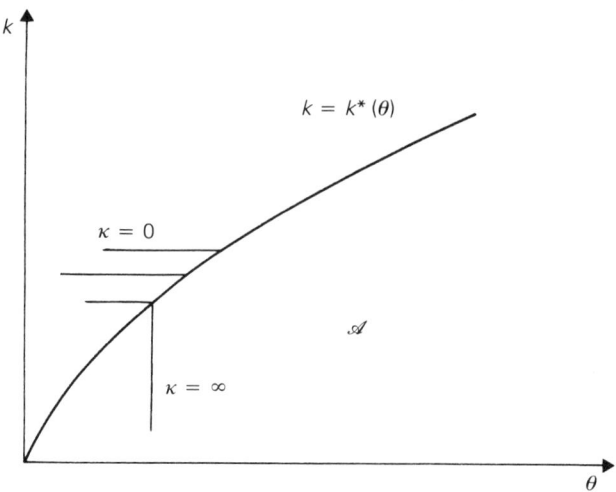

Fig. 10.2 The optimal investment strategy.

as the reservation level for Θ_t at the capacity level $K_t = k$. As Θ_t fluctuates to the left of the curve, nothing will happen until Θ_t again hits the boundary $\theta^*(k)$. Then an infinitesimal amount of capacity will be invested, and the process is again deflected to the left. As long as Θ_t is to the left of the curve, demand is so low that it is not worthwhile to expand the capacity.

The process cannot stay in the area $\mathscr{A}$ for any finite period of time. Hence, investment only takes place along the curve $k^*(\theta)$. This is the basic structure of the singular solution of a problem of irreversible investment. In our model, the solution is complicated by the fact that the electricity price is either determined by the equilibrium condition or by the gas price, as expressed by Equation 10.13. If we ask how high θ has to be before demand is higher than the given hydro-power capacity and thermal power is phased in, we find from Equation 10.13 that $(\theta/k)^{1/\varepsilon} = q$, or equivalently, $k = q^{-\varepsilon}\theta$. This corresponds to a ray in the (θ, k)-plane, which is the boundary between the regions (A) $k > q^{-\varepsilon}\theta$, and (B) $k \leq q^{-\varepsilon}\theta$ (Fig. 10.3a). In (A) only hydro power is used, and the electricity price is lower than q. In (B) thermal power is used, and the electricity price is equal to q. The complete solution thus involves both the singularity of the investment rate and the two different pricing policies in regions (A) and (B).

We will now briefly sketch the mathematical solution of the stochastic control problem. The solution can be found by applying a generalized version of the so-called Hamilton–Jacobi–Bellman (HJB) equation. In our case the HJB equation corresponding to Equation 10.12 is

$$\max_{\kappa} \left\{ (f(\theta, k) - C(k)\kappa)e^{-rt} + \frac{\partial H}{\partial t} + \kappa \frac{\partial H}{\partial k} + \alpha\theta \frac{\partial H}{\partial \theta} + \tfrac{1}{2}(\beta\theta)^2 \frac{\partial^2 H}{\partial \theta^2} \right\} \leq 0 \tag{10.14}$$

where equality is required only for $(\Theta_t, K_t) \notin \mathcal{A}$. Here $f(\theta, k)$ is consumer surplus less cost of gas for thermal power production. With the specified demand function $f(\theta, k)$ is

$$f(\theta, k) = \begin{cases} \dfrac{\varepsilon}{\varepsilon - 1} \left(\dfrac{\theta}{\kappa}\right)^{1/\varepsilon} k & \text{for } k > q^{-\varepsilon}\theta \\ qk + \dfrac{1}{\varepsilon - 1} \theta q^{1-\varepsilon} & \text{for } k \leq q^{-\varepsilon}\theta \end{cases} \tag{10.15}$$

Strictly speaking, the given solution for $f(\theta, k)$ applies only when $a = 0$ and $\varepsilon > 1$. The functional form of $f(\theta, k)$, in the case of $\varepsilon < 1$, includes a constant term, depending on $a > 0$, which is irrelevant for the optimal investment strategy, and we can therefore apply the solution also for the case of $\varepsilon < 1$.

We will omit the mathematical details of the solution to Equation 10.14 and, rather, elaborate the graphical discussion begun above. In Fig. 10.3(a) we have repeated the curve $k = k^*(\theta)$ from Equation 10.2, and included the ray $k = q^{-\varepsilon}\theta$, which is the boundary between the two regions of different pricing policies. Moreover, the solid curve k_c shows the optimal capacity under full certainty. We see that for a given θ, optimal capacity is lower under uncertainty, which is a consequence of the irreversibility of investment. In Fig. 10.3(b) we have replaced θ on the abscissa by $p = (\theta/k)^{1/\varepsilon}$, which represents the equilibrium price if $p \leq q$. In this case the boundary between regions (A) and (B) becomes vertical. We have defined θ^* as the reservation level for θ_t at the capacity level $K_t = k$. Correspondingly, we define the reservation price $p^* = (\theta^*/k)^{1/\varepsilon}$. The figure shows that the reservation price must be much higher under uncertainty than under certainty to warrant investment of the same amount of capital in hydro power.

The line $k^{\max}$ represents the upper limit for future expansion of the hydro-power system, and it is given by

$$C(k^{\max}) = \frac{q}{r} \tag{10.16}$$

Under full certainty, hydro-power capacity is expanded for increasing θ until the point where the electricity price reaches q. From then on, capacity is constant at $k^{\max}$. The interpretation of $k^{\max}$ is therefore the maximal capacity under full certainty. Under uncertainty, the capacity will never reach $k^{\max}$, but approach this level asymptotically as $\theta \to \infty$.

If the decision maker were in $\mathcal{A}$, the optimal decision would be to invest immediately to reach $k^*(\theta)$. Starting from a point outside $\mathcal{A}$, however, all

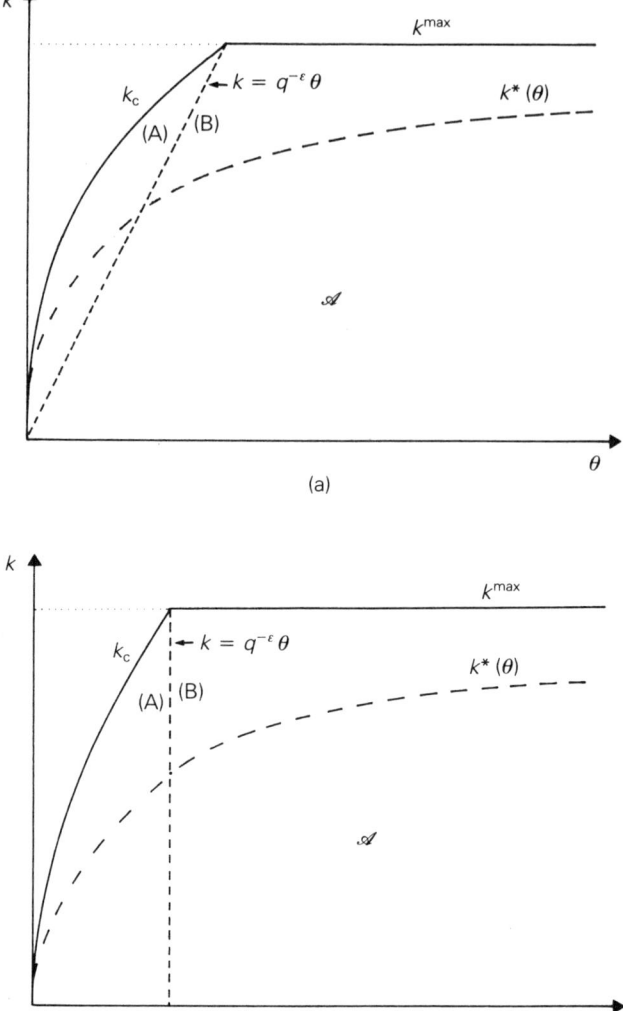

Fig. 10.3 Optimal capital stock K_t^* in hydro power: (a) as function of θ; (b) as function of P.

investment would take place along $k^*(\theta)$. In fact, by acting optimally we would never cross into $\mathcal{A}$. In order to illustrate the time dimension of the investment process, Fig. 10.4 gives K_t^* and a sample path of the capital stock K_t as time functions. Hydro-power investments are undertaken only when $K_t^* \geq K_t$, and K_t is constant on the intervals where $K_t^* < K_t$.

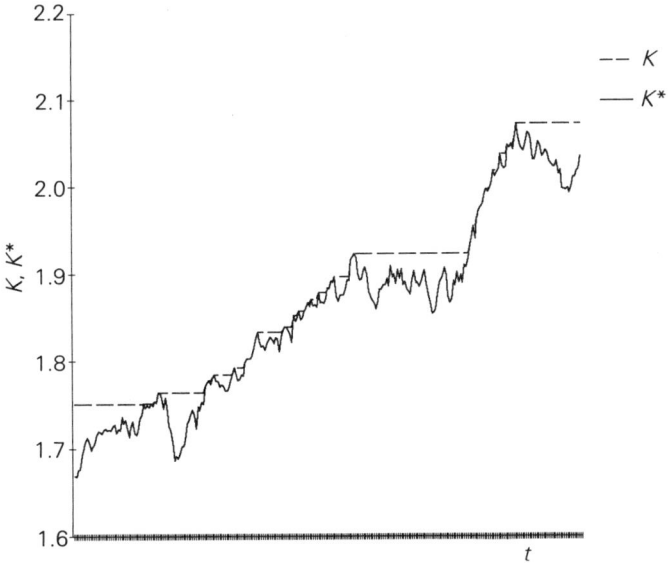

Fig. 10.4 Optimal (K_t^*) and actual (K_t) capital stock in hydro power.

Moreover, the initial capacity K_0 is a lower bound for the optimal capacity at any time. Formally, the relationship between actual and optimal capacity is given by

$$K_t = \max\{K_0, \max\{K_s^*: s \leq t\}\}$$

Solving the HJB equation gives the expression for the curve $k^*(\theta)$. It is more convenient to work with the inverse $\theta^* = \theta^*(k)$. In region (A) of Fig. 10.3, $\theta^*(k)$ is given by the equation

$$\theta^*(k) = \theta_A^*(k) = \left[\frac{r(\varepsilon - 1/\gamma)}{\varepsilon} C(k)\right]^\varepsilon k \quad \text{for } k < k^{\max} \quad (10.17)$$

In region (B) of Fig. 10.3, that is, when price equals fuel costs q, we have

$$\theta^*(k) = \theta_B^*(k) = \left[\frac{r(1 - \varepsilon\gamma)}{q}\left(\frac{q}{r} - C(k)\right)\right]^{1/\gamma} kq^\varepsilon \quad \text{for } k < k^{\max} \quad (10.18)$$

where γ is the negative root of the characteristic equation from the HJB equation

$$\tfrac{1}{2}\beta^2\gamma^2 + (\alpha - \tfrac{1}{2}\beta^2)\gamma - r = 0 \quad (10.19)$$

The curve $\theta^*(k)$ is continuously differentiable at the boundary between (A) and (B). Note that in region (A) the expression for $\theta^*(k)$ does not depend upon q, since the gas price is of no relevance for the investment decision as long as $P < q$.

200 The choice between hydro and thermal power generation

Figures 10.5(a–d) show the effect on the reservation price P^* of four key parameters: the demand elasticity (ε), the demand drift (α), the demand variability (β), and the gas price (q). Figure 10.5(a) shows that the reservation price is lowered by a higher demand elasticity. The disadvantage of premature irreversible investment is mitigated by flexible demand and the reservation price can be lowered. Figure 10.5(b) shows that the reservation price increases sharply with the degree of uncertainty when the gas price is close to the reservation price at $\beta = 0$ (dashed curve), while it is influenced very little when the gas price is high (solid curve). For small values of β the two curves are indistinguishable and quite flat, reflecting that gas is unlikely to be used in the short run.

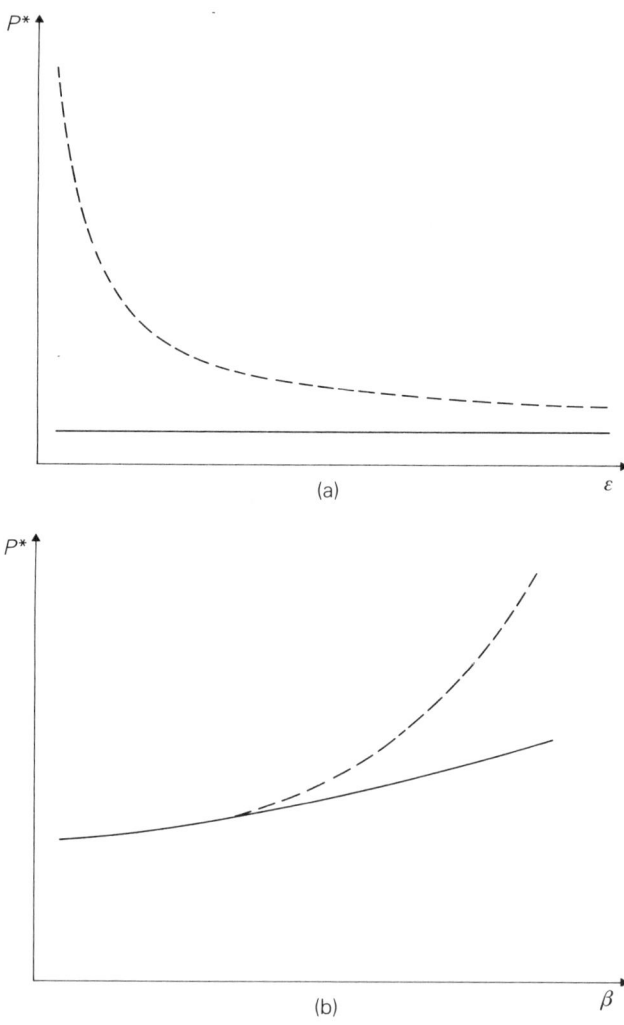

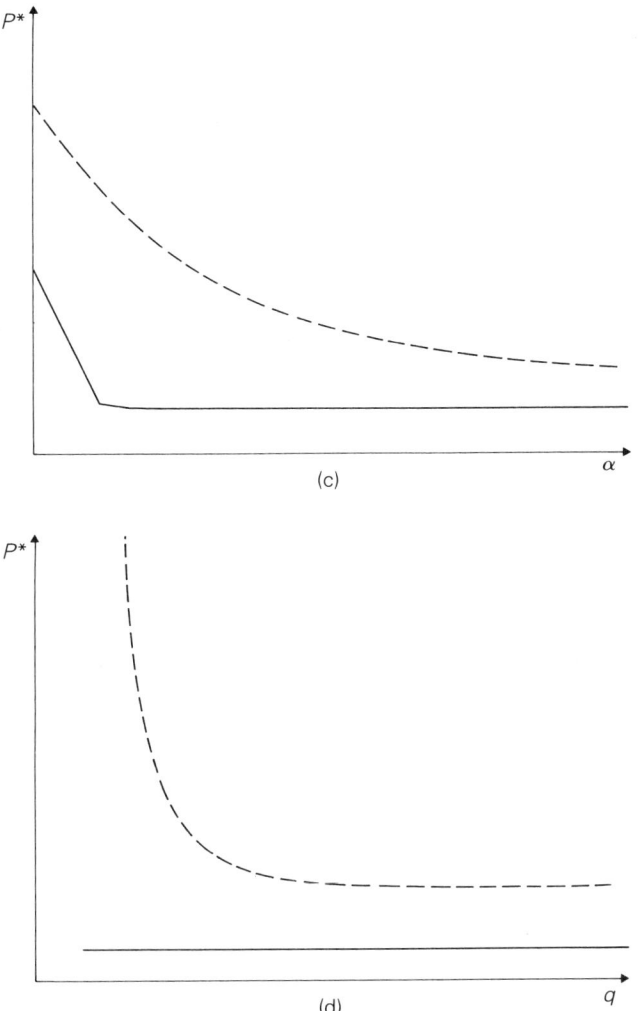

Fig. 10.5 The reservation price for different value of the parameter: (a) as a function of demand elasticity under certainty and uncertainty; (b) as a function of standard deviation β with $q = 0.7$ and $q = \infty$; (c) as a function of the drift parameter α under certainty and uncertainty; (d) as a function of the gas price under certainty and uncertainty.

Figure 10.5(c) shows the effect of the drift parameter under certainty and under uncertainty. When the drift parameter is non-negative, the reservation price is constant under certainty. How fast demand grows should thus not influence when to effectuate new investment. If the drift is negative,

202 The choice between hydro and thermal power generation

however, a higher reservation price is required. Under uncertainty the kinked curve is smoothing out with the maximal distance between the curves at $\alpha = 0$.

Figure 10.5(d) shows the dependence of the reservation price upon the gas price. For a given capacity k the reservation price is independent of the gas price for high values of q. For the gas price sufficiently low, the reservation price increases sharply, since it is not worthwhile to expand the hydro-power capacity as the thermal alternative becomes sufficiently cheap.

Details of the solution can be found in Kobila (1989), a more elaborate mathematical version of this chapter.

10.4 FINAL REMARKS

The preceding sections have focused entirely on specific aspects of the choice between hydro and thermal power generation. The setting of the problem has been highly simplified. The purpose has been to highlight features that tend to be subdued in current government planning.

Among the simplifying assumptions made here are the stylized description of the cost structure with only the fuel cost of thermal power taken into consideration, and the cost of hydro power represented as everlasting capital. Furthermore, in section 10.3 hydro power is dealt with as if it could be infinitesimally expanded. In fact, hydro-power plants are large units that take a long time to approve, plan and build, say 5–10 years. The construction lag does not change, however, the criteria developed above for choosing between a hydro plant and thermal power. The indivisibility of hydro plants changes the problem in a more substantial way.

On the gas side there are also a number of issues that makes the treatment above stand out as highly simplified. The opportunity cost of using gas for electricity generation is not well defined. Some actual gas fields will be unconnected to a market grid in the foreseeable future, while others produce associated gas that would otherwise be flared (if flaring was allowed). Oligopolistic market elements also contribute to an opportunity cost below the export price.

There are some important aspects of the interaction between hydro power and thermal power that deserve to be mentioned and should be taken into consideration in a more fully developed planning model. Hydro power has its own uncertainty caused by variations in rainfall over the year and between years. The normal seasonal variations are considerable, implying that the marginal cost of hydro power varies over the year. This implies that an alternative source of supply with low capital cost may have a role to play in an integrated supply system even when its marginal cost is higher than the annual average marginal cost of hydro power. The variations in marginal cost of hydro power over the year are exacerbated by seasonal demand variations.

The variations in hydro-power supply between years is also considerable and is countered in a pure hydro-power system by large and expensive reservoirs. This provides a premium for a thermal source that has not been taken into consideration above. These uncertainty aspects are somewhat more complicated to deal with analytically than those studied in this chapter.

APPENDIX

Solution of the optimal stopping problem

The solution to the optimal stopping problem is found by the following theorem:

THEOREM 1. *Let*

$$V(t, q; \kappa, r) = \max_{\tau} E^{t,q}\{(Q_\tau - \kappa)e^{-r\tau}\} \quad (A.1)$$

where

$$dQ_s = \mu Q_s ds + \sigma Q_s dB_s \quad (A.2)$$

Then

$$V(t, q; \kappa, r) = \frac{1}{\gamma - 1}(q^*)^{-\gamma} \kappa q^\gamma e^{-rt} \quad (A.3)$$

where the reservation price

$$q^* = \frac{\gamma}{\gamma - 1} \kappa$$

with

$$\gamma = \frac{1}{\sigma^2}[-(\mu - \tfrac{1}{2}\sigma^2) + \sqrt{\{(\mu - \tfrac{1}{2}\sigma^2)^2 + 2r\sigma^2\}}] > 1 \quad (A.4)$$

The optimal stopping rule is

$$\tau = \inf\{t > 0: Q_t > q^*\} \quad (A.5)$$

or, simply, to wait until Q_t is equal to the reservation price.

Theorem 1 is a well-known result and a special case of the problem solved by McDonald and Siegel (1986). We apply the theorem by using $\kappa = (r_Q - \mu)K$.

In order to show that the reservation price is higher under uncertainty, we need to consider an alternative expression for γ. Define $\hat{\mu}$ by

$$\hat{\mu} = \mu - (r_Q - r_F) \qquad (A.6)$$

It follows from the proof of the theorem that γ can be expressed in terms of the risk-free discount rate r_F, rather than in terms of the option value discount rate r, provided that μ is replaced by the adjusted drift parameter $\hat{\mu}$. By substituting the parameters in Equation A.4 we get

$$\gamma = \frac{1}{\sigma^2}\left[-(\hat{\mu} - \tfrac{1}{2}\sigma^2) + \sqrt{\{(\hat{\mu} - \tfrac{1}{2}\sigma^2)^2 + 2r_F\sigma^2\}}\right] > 1 \qquad (A.7)$$

It follows that

$$\frac{\gamma}{\gamma - 1} > \frac{r_F}{r_F - \hat{\mu}}$$

and since $r_Q - \mu = r_F - \hat{\mu}$, the reservation price q^* exceeds $r_F K$.

ACKNOWLEDGEMENTS

The authors wish to thank David Newbery, Øystein Olsen and Diderik Lund for helpful suggestions and comments.

REFERENCES

Arrow, K. J. and Fisher, A. C. (1974) Environmental preservation, uncertainty and irreversibility, *Quarterly Journal of Economics*, 88, 312–19.

Bjerkholt, O. and Brekke, K. A. (1988) Optimal starting and stopping rules for resource depletion when price is exogenous and stochastic. Discussion Paper from the Central Bureau of Statistics, Oslo, no. 40.

Brennan, M. J. and Schwartz, E. S. (1985) Evaluating natural resource investments. *Journal of Business*, 58, 135–57.

Ekern, S. (1988) An option pricing approach to evaluating petroleum projects. *Energy Economics*, 10, 91–9.

Fisher, A. C. and Hanemann, M. (1982) Valuing pollution control: the hysteresis phenomenon in aquatic ecosystems, Working Paper, University of California, Berkeley.

Kobila, T. Ø. (1989) *An Application of Reflected Diffusions to the Problem of Choosing between Hydro and Thermal Power Generation*, Reprint Series, no. 5, 1989, Institute of Mathematics, University of Oslo.

McDonald, R. and Siegel, D. (1986) The value of waiting to invest. *Quarterly Journal of Economics*, 101, 707–27.

Majd, S. and Pindyck, R. S. (1985) Time to build, option value and investment decisions. Working Paper, Energy Laboratory, MIT.

Massé, P. (1962) *Optimal Investment Decisions: Rules for Action and Criteria for Choice*, Prentice-Hall, Englewood Cliffs, New Jersey.

Øksendal, B. (1985) *Stochastic Differential Equations: An Introduction with Applications*, Springer-Verlag, Heidelberg.

Olsen, T. E. and Stensland, G. (1987) Optimal sequencing of resource pools under uncertainty. Unpublished paper.

Olsen, T. E. and Stensland, G. (1988) Optimal shut down decisions in resource extraction. *Economic Letters*, **26**, 215–18.

Paddock, J. L., Siegel, D. R. and Smith, J. L. (1988) Option valuation of claims on real assets: the case of offshore petroleum leases. *Quarterly Journal of Economics*, **102**, 497–508.

Pindyck, R. S. (1988) Irreversible investments, capacity choice, and the value of the firm. *American Economic Review*, **78**, 969–85.

Roberts, K. and Weitzman, M. L. (1981): Funding criteria for research, development and exploration projects. *Econometrica*, **49**, 1261–88.

Shiryayev, A. N., (1978) *Optimal Stopping Rules*, Springer-Verlag, Heidelberg.

11

The management of jointly produced exhaustible resources

JON VISLIE

11.1 INTRODUCTION

Until recently most contributions dealing with extraction of exhaustible resources have analysed problems related to the management of a single pool of a resource, such as oil. There are, however, many types of resources that are discovered and extracted jointly, like natural gas and oil. Pindyck (1982) discusses optimal management of jointly produced exhaustible resources; the market behaviour for resources that are extracted jointly is analysed, and it is demonstrated how the price of one resource will depend upon the demand and storage cost for the other. Abodunde and Wirl (1985) consider a country (or a resource cartel like OPEC) extracting both oil and natural gas (associated gas and gas from dry basins), and facing downward-sloping demand curves for both resources. They derive an optimal extraction programme and find various combinations of oil and gas extraction. One interesting result is that the country might find if profitable to flare an amount of gas in order to keep the oil price at a desired level. Wirl (1987) has a similar model, with only associated gas, where the main result is that wasting (i.e. flaring the gas) could be optimal from the producer's point of view, at least over a period of time, to increase the revenue from the other commodity. (Whether flaring will be total or partial is shown by Wirl to depend upon the degree of substitutability in demand.)

None of these papers capture the main features of the market for natural gas in Western Europe. This market is characterized by a small number of exporting countries, like the USSR, Norway and Algeria, and a few large importing countries, like West Germany, France, Italy and the UK. The market structure is somewhere between monopoly and perfect competition, and we could name it **bilateral oligopoly**. In this market natural gas is usually traded according to long-term contracts established in negotiations between the parties involved, usually large companies, some representing national interests, while others are (more or less regulated) profit-maximizing monopolies. The main reason why natural gas is traded according to long-term contracts is that there exists no spot market for natural gas as trade is made possible only after the parties have built up installations, such as pipelines, terminals and other transportation facilities. As most of these investments are relationship-specific and irreversible, the parties are locked into a bilateral monopoly position once the investments have been undertaken. The long-term contract will therefore serve as a guarantee for either party, such that neither of them will be 'trapped' by the other party at a later point in time.

There is no international jurisdiction to ensure that nations honour contractual agreements. The terms of the contract must then be determined such that neither party will find it profitable at any point in time to break the agreement; hence the contract is required to be self-enforcing. In deriving this long-term contract, we assume that the parties have perfect foresight and can enter into a complete contract for the entire period of their trading relationship. This approach is in contrast to that proposed by Hart and Moore (1988), who formulate a model where the parties will seldom be able to anticipate all contingencies that may take place during the relationship. They will usually engage in an incomplete contract, which specifies a procedure for revising the contractual provisions as new information is revealed.

The problem of joint production (i.e. that oil and natural gas are produced in fixed proportions) seems to be relevant for some fields in the Norwegian sector of the North Sea. It should therefore be of some interest to examine how such long-term contracts for natural gas are established and the impact on the terms of these contracts of natural gas being produced jointly with oil. In order to reach tractable conclusions, we will assume that there is only one seller and one buyer of natural gas. We assume, futhermore, that the long-term contract for natural gas is established as a Nash bargaining solution, and that oil is sold in a competitive world market. The latter assumption seems to be realistic from the perspective of Norway which has a negligible market power in the oil market, but is considered as a 'large' seller of natural gas to the European continent. (Hoel and Vislie (1987) derive a long-term contract for natural gas, established as a Nash bargaining solution, in a bilateral context, but in the absence of any joint production.)

In Vislie (1986) a related problem is examined in a static framework, with jointness not as a technological characteristic, but a consequence of a multiproduct cost function (economies of scope), and with a decision sequence opposite to that chosen in this chapter. Instead of first choosing contract rules of the gas contract, as in Vislie (1986), we will now consider a situation where the seller of oil and gas first decides on a decision rule for selling oil in a competitive world market, and then negotiates with the buyer about a gas contract. The main problem addressed is how this long-term contract for natural gas is determined when natural gas is extracted jointly with oil, and in what way gas deliveries and oil extraction are mutually affected by this production relationship.

The setting is then as follows: a resource-rich country negotiates with a potential buying country, represented by a national enterprise, about a long-term contract for natural gas delivery for an indefinite period. In order to implement the contract, both parties will usually have to undertake specific investments, which will normally take time to be completed. However, in order to emphasize the aspects of joint production, we will ignore the problems related to investments in infrastructure. Hence, we will not enter into any discussion about whether or not such specific investments will be efficient; see e.g. Grout (1984) for a related problem.

11.2 THE STRUCTURE OF THE MODEL

Natural gas is traded between one seller and one buyer in a long-term contract that is established as a Nash bargaining solution. The seller is a profit-maximizing company, whereas the buyer is a public enterprise, representing the interests of the ultimate consumers of natural gas in the importing country. Oil, produced jointly with natural gas, can always be sold in a competitive world market and the seller is assumed to have perfect foresight about the future price path of oil.

If a long-term contract for natural gas is established (at the beginning of the period under consideration), specifying that deliveries will start at $t = 0$ and lasting indefinitely, the buyer's net surplus will be equal to the present value of consumers' surplus:

$$V = \int_0^\infty e^{-rt} [U(y(t)) - p(t)y(t)] dt \qquad (11.1)$$

where $U(y(t))$ is the buyer's instantaneous utility in some monetary unit, with $y(t)$ as the volume of natural gas delivered at t, $p(t)$ the price of natural gas and r a constant international interest rate. Here U is assumed differentiable, strictly increasing and strictly concave in y. (It should be noted that the consumers' gross benefit function $U(y)$ is expressed only as a function of gas consumption. This does not imply that prices of energy substitutes, like oil, electricity and coal, have no impact on the demand for natural gas. The formulation in Equation 11.1 can be justified on the grounds that the consumers have perfect foresight about the future prices of these substitutes; so $U(y)$ is the gross benefit of gas, contingent upon the future prices of energy substitutes.)

On the other side of the market we have the seller of both oil and natural gas. His total pay-off or present value of net revenues from selling oil (W_o) and natural gas (W_n) will be

$$W \equiv W_o + W_n = \int_0^\infty e^{-rt} \pi(x(t))\,dt + \int_0^\infty e^{-rt}[p(t) - v]y(t)\,dt \quad (11.2)$$

where $\pi(x) \equiv qx - C(x)$ is the current net revenue from selling oil volume x at price q, π is assumed to be strictly concave in x with $\pi'(0) > 0$, implying that some production of oil is profitable at any oil price and v is a constant unit transportation cost for natural gas.

In order to formulate the bargaining problem properly, we should first derive the selling country's optimal oil policy, for any admissible gas contract, represented by an arbitrary delivery path $y(\cdot)$ in $[0, \infty)$. The optimal oil programme, contingent upon a delivery profile for natural gas, is found as the solution to the following problem:

$$\underset{x(\cdot)}{\text{maximize}} \int_0^\infty e^{-rt} \pi(x(t))\,dt \quad (11.3)$$

$$\text{s.t.} \int_0^\infty x(t)\,dt \leq S_0 \quad \text{and} \quad z = kx \equiv x \geq y \geq 0 \quad \text{for all } t$$

where S_0 is the initial stock of oil and z, which is production of natural gas, is proportional to oil extraction at any t. We furthermore assume that the coefficient $k \equiv 1$, by proper normalization of the units of measurement. The constraints in Equation 11.3 specify, first, that total accumulated oil extraction cannot exceed the initial stock of oil reserves. Secondly, in order to meet any gas delivery at any point in time, the production of associated gas at t, $z(t)$, which is proportional to oil production at t, must at least be equal to the delivery requirement at t. Let Y denote any admissible delivery path for gas, and let $W_o^*(Y)$ be the maximal value of the objective function in Equation 11.3 given the constraints. This value function can be derived using ordinary control theory. Let $\alpha(t)$ be the adjoint function associated with the resource constraint in Equation 11.3 and $m(t)$ be a non-negative shadow price associated with the gas delivery requirement. The Lagrangian function for the problem in Equation 11.3 is then

$$L = e^{-rt}\pi(x) - \alpha x + m(x - y) \quad (11.4)$$

Necessary conditions for $x^*(\cdot)$ to be the extraction path for oil that solves the problem in Equation 11.3 are then

$$e^{-rt} \cdot \pi'(x^*(t)) - \alpha(t) + m(t) = 0 \quad (11.5)$$

$$\dot{\alpha}(t) = -L'_S = 0 \Rightarrow \alpha(t) = \alpha \geq 0 \ (= 0 \text{ if } \lim_{t \to \infty} S(t) > 0)$$

$$m(t) \geq 0 \ (= 0 \text{ if } x^*(t) > y(t)) \quad (11.5)$$

where $S(t)$ is defined as the remaining oil reserves at t. (As the state variable $S(t)$ does not enter the Lagrangian function, α is a non-negative time-independent constant, but will of course depend on the terms of the gas contract Y, that is $\alpha = \alpha(Y)$.)

The first condition in Equation 11.5 is the ordinary rule for optimal extraction, stating that the present value of marginal profit from extracting oil should be equal to the shadow price of oil, α, adjusted for a shadow price associated with the natural gas requirement. (Note that $m(t)$, which is non-negative, will be equal to minus the partial derivative of the value function $W_o^*(Y)$ with respect to $y(t)$, indicating the marginal loss in the maximal value from selling oil due to a higher gas delivery $y(t)$ at t.)

Let us now assume that the oil reserves, in the absence of any gas contract, i.e. $Y = 0$ or $y(t) = 0$ for all t, will be depleted in finite time, so $\alpha(0)$ will be positive. In that case any gas delivery requirement Y will imply that $\alpha(Y) \geq \alpha(0)$, with strict inequality if the shadow price m is positive for some t, which means that the optimal extraction path for oil has to be totally adjusted in order to meet the gas requirement for some t. Hence, the optimal oil extraction plan, contingent upon some gas delivery profile Y, can be expressed as $x^* = x^*(Y)$, with an extraction at t equal to $x^*(t, Y)$. Properties of this function, along with the value function $W_o^*(Y)$, will play a crucial role for the determination of the terms of the gas contract, to which we now turn.

11.3 THE DERIVATION OF THE TRADING RULES OF THE GAS CONTRACT

In this section we will derive the terms of the trading rules for natural gas, as well as the associated optimal extraction programme for oil. The natural gas contract is, if an agreement is reached, determined as a Nash bargaining solution. This contract will in the present context stipulate a delivery profile and a payment schedule. If, however, agreement is not being reached, the seller will sell only oil in the world market, while the production of natural gas is flared, leaving the seller with a pay-off equal to $W_o^*(0)$, which is the maximal present value of selling oil in the absence of a gas contract. This pay-off will be the seller's **disagreement point** in the formulation of the bargaining problem. The buyer, on the other hand, has by assumption a disagreement point equal to zero.

The long-term contract for natural gas is found by maximizing the product of the agents' pay-offs in excess of their disagreement points:

$$\underset{y(\cdot), p(\cdot)}{\text{Maximize}} \quad \{V[W - W_o^*(0)]\}$$

s.t. $V \geq 0 \qquad W \geq W_o^*(0) \quad \text{and} \quad y(t) \leq x^*(t, Y) \quad \text{for all } t$

(11.6)

212 The management of jointly produced exhaustible resources

The objective function in Equation 11.6 is the Nash product for the problem at hand; the first two constraints are the 'participation' or individual rationality constraints, whereas the last one represents the constraint saying that gas delivery at t cannot exceed oil production at t, where oil production is determined according to Equation 11.5. (Note that the overall resource constraint in Equation 11.3 is satisfied with the quantity constraint in Equation 11.6. Hence, if there exists a delivery path for natural gas that solves the problem in Equation 11.6, this gas delivery profile can in fact be realized.)

As the agents use the same discount factor, the bargaining between the seller and the buyer can be broken down into two steps: first, choose a delivery path for natural gas, admissible by the quantity constraint in Equation 11.6, that maximizes total surplus (or yields the greatest possible 'pie') from the agreement, as given by $V + W_n + W_o^*(Y)$. Second, choose a payment schedule or a price path such that the maximal total surplus is divided according to what the solution of Equation 11.6 prescribes. (As will be revealed later, a unique price path cannot be derived directly from our solution. However, in order to reach a contract stipulating a unique price at each point in time, we will introduce an additional requirement, saying that the price path is to be dynamically consistent. Within the present framework, one could, even though the term is used differently in the game-theorectic literature, say that the requirement of dynamic consistency represents 'renegotiation-proofness'.)

11.3.1 The delivery path for natural gas

The surplus-maximizing delivery path for natural gas, denoted Y^*, is found as the solution to the following problem:

$$\underset{y(\cdot)}{\text{Maximize}} \; V + W_o^*(Y) + W_n \quad \text{s.t.} \; y(t) \leq x^*(t, Y) \quad \text{for all } t \quad (11.7)$$

Let us now suppose that the buyer's marginal willingness to pay, evaluated for $y = 0$, $U'(0)$, is strictly greater than the marginal transporation cost v. Let the value of y that satisfies the condition $U'(y) = v$, be denoted $\bar{y}$, which is positive. Let us furthermore suppose that a delivery path with $\bar{y}$ at each point in time is not feasible. The optimal delivery path for natural gas is then found from the following condition:

$$U'(y^*(t)) - v - e^{rt} m(t) = 0 \quad \text{for all } t \in [0, \infty) \quad (11.8)$$

stating that for the gas delivery that maximizes the total surplus or 'pie', the buyer's current marginal gross valuation is equal to the marginal transportation cost plus a term reflecting the seller's loss in current net revenue from selling oil, due to the gas contract. As $m(t)$ is non-negative, (cf. Equation 11.5), the gas delivery at any t where the constraint in Equation 11.7 is binding, must be adjusted downwards as compared to $\bar{y}$. Since the delivery path $\{\bar{y}(\cdot)\}$ is not feasible during the entire contracting period, there must exist some time interval in $[0, \infty)$ during which the shadow price m will be positive.

As long as oil production is sufficiently high to meet the gas requirement, we have $y^*(t) = \bar{y}$. On the other hand, when oil production is no longer sufficiently high to realize a gas delivery equal to $\bar{y}$, gas delivery will be smaller. The fact that the quantity constraint in Equation 11.7 will be binding during some time interval, will also cause the seller to adjust downwards his entire extraction path of oil, as we, with our assumptions, will have $\alpha(Y^*) > \alpha(0)$. Returning to the conditions for optimal oil extraction in Equation 11.5, we observe that as $\alpha(Y^*) > \alpha(0)$, the optimal oil path will be adjusted downwards during time intervals where x^* is higher than y^*, i.e. when $\bar{y} = y^*$.

We therefore have that when the quantity constraint in Equation 11.7 is not binding, gas delivery will be equal to $\bar{y}$, as seen from Equation 11.8 with $m = 0$. In time intervals during which the quantity constraint is binding, i.e. when $x^* = y^*$, gas delivery will be smaller than $\bar{y}$, and determined according to what is being stipulated in Equation 11.9, where we have used the first condition in Equation 11.5 in order to eliminate m.

$$U'(y^*(t)) + q(t) = v + C'(x^*(t, Y^*)) + e^{rt}\alpha(Y^*) \tag{11.9}$$

This optimality condition can be interpreted as follows: on increasing gas delivery at t, when $x(t) = y(t)$, the increment in total current gain or surplus is $U'(y(t)) + q(t)$; the sum of the buyer's marginal gross benefit from gas consumption and the seller's marginal revenue from selling oil. The current increase in cost of providing this marginal unit of gas is $v + C'(x)$; the sum of marginal transportation and extraction costs, and a term αe^{rt} which is the current marginal cost of reducing the stock of oil at t, reflecting the scarcity value of oil in the ground. The term αe^{rt} therefore reflects the current marginal cost to the seller from not being able to pursue the first-best marketing policy for oil, as given by the oil extraction plan leading to $W_o^*(0)$. We then have that this way of determining the gas delivery profile implies a trade-off between the marginal gains and losses to either party, when the quantity constraint is binding. The rule in Equation 11.9 can then be interpreted as one where each party makes some concession to his opposite number as both the oil programme and the delivery profile for natural gas are being adjusted so as to maximize total surplus, $V + W$, to the agents. In Fig. 11.1 we have illustrated how the extraction path for oil, with no gas contract, $x^0(\cdot)$, differs from the one following from a long-term delivery contract for natural gas. We have assumed that the stock of oil is depleted at date T, which is the same in both contexts. (Note that in our problem discussed above, there is no reason to believe that the date of depletion should be the same in the two

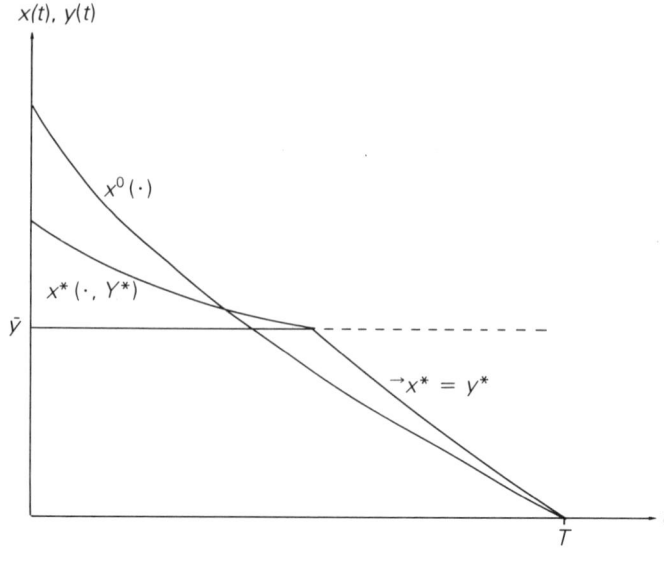

Fig. 11.1

cases. When taking this plausible fact into account, the two extraction paths would have differed even more.)

11.3.2 The price path

The second step of our problem consists of deriving the price path. Given the optimal gas delivery profile, which was found to be independent of the price path, the optimal price path must satisfy the condition of equal sharing of the net surplus, that is, the pay-off in excess of the disagreement point should be the same for both agents. (We have assumed that the agents have equal bargaining power, as we have used the symmetric Nash bargaining solution.) The price path must therefore satisfy the following condition for dividing the surplus between the two:

$$V(Y^*) = W_n(Y^*) + W_o^*(Y^*) - W_o^*(0) \qquad (11.10)$$

From this, however, we cannot derive a unique price path. In fact a continuum of price paths will satisfy the equilibrium condition 11.10. In order to reach a unique solution, we will require the price path to be time consistent. This requirement, which is outlined for a related problem in Hoel and Vislie (1987) and in Vislie (1987), and discussed more generally by Newbery (1981) for extraction of exhaustible resources, is a natural requirement for a true equilibrium payment schedule, as this requirement can be seen as capturing the idea of a self-enforcing contract. As agents (here: national enterprises) involved in international transactions will not sign binding contracts, the only type of contract that will be honoured within such a setting is one that will be enforced by mutual self-interest. In order to determine a unique price path we must look for a rational expectations equilibrium, where the price path is maximizing the Nash product not only at $t = 0$ but for any $t \in [0,\infty)$. This means that the equilibrium price path can be determined as if renegotiations took place continuously from the present ($t = 0$) and onwards. Suppose that the parties renegotiated at $\tau \in (0,\infty)$ about the terms of the contract. In that case the symmetric Nash bargaining solution for the period $[\tau,\infty)$ would stipulate terms that are identical to the provisions of the original long-term contract. In order to determine this dynamically consistent price path, let us introduce the following definition:

$$\mu(t) \equiv \pi(x^0(t)) - \pi(x^*(t, Y^*)) \qquad (11.11)$$

that is, the difference between the current oil revenue at t from pursuing the optimal extraction plan with and without a gas contract. Hence $\mu(t)$ is the current loss in net oil revenue per unit of time due to the gas contract. Suppose that after having pursued some part of their trading relationship, the parties were offered the opportunity to revise their original long-term contract, say at $t = \tau$. According to what has been said above, the price path from τ and onwards should then be as follows:

$$\int_\tau^\infty e^{-rt} p(t) y^*(t) dt = \frac{1}{2} \int_\tau^\infty e^{-rt} \{U(y^*(t)) + vy^*(t) + \mu(t)\} dt \qquad (11.12)$$

Since this condition must hold for any $\tau \in [0, \infty)$, we can derive a unique price at each point in time during the contract period, as given by

$$p^*(t) = \frac{1}{2} \left\{ \frac{U(y^*(t)) + \mu(t)}{y^*(t)} + v \right\} \qquad (11.13)$$

The interpretation of Equation 11.13 is as follows: along a time consistent price path, the unit price at each moment of time during the contract period is set equal to the arithmetic average of the buyer's average gross benefit and the seller's unit costs per unit of time. (The unit costs of the seller is given by the sum of unit transportation cost v, and the current loss in net oil revenue at t due to the gas contract per unit gas delivered at t.) In formulating such a pricing rule, the seller and the buyer will cover their unit costs, and in such a way that the current net surplus, adjusted for the current loss in net oil revenue due to the gas contract, is divided equally between the two parties.

11.4 CONCLUSIONS

In this chapter we have analysed oil extraction and long-term contracts for natural gas within a framework of strict joint production between oil and natural gas, when oil is always being sold in a competitive world market and natural gas, if agreement is being reached, is traded according to a long-term contract determined as a Nash bargaining solution. As noted above, the optimal extraction plan for oil, in the presence of a gas contract, will be altered as compared to the extraction plan that would have been optimal if no gas contract were established. The reason is that, with our assumptions, the gas contract will impose a binding constraint on the extraction of oil during some time interval. Hence, as illustrated in Fig. 11.1, the oil extraction will shift from $x^0(\cdot)$ to $x^*(\cdot, Y^*)$. For the same reason, the delivery profile for natural gas will also be adjusted, as a delivery per unit of time $\bar{y}$ is not feasible. This interactive adjustment is worth noting, as it illustrates an interesting feature: that the optimal oil programme can be significantly altered in the presence of joint production, when the joint product (here natural gas) is marketed according to a long-term contract.

We also derived an equilibrium payment schedule for how the net surplus from natural gas trade should be divided between the two parties. This payment schedule or price path was required to be dynamically consistent, as this was assumed to capture the idea of a self-enforcing contract. As demonstrated above, the price of natural gas at any point in time compensates the seller for the current loss in net oil revenue from not being able to pursue the 'first-best' oil programme due to the gas contract.

Lastly, let us compare some of the results emerging from the present model with some of the results derived by Wirl (1987). He also analyses the question about joint production and exhaustible resources, but instead of introducing long-term contracts for natural gas, Wirl considers a country or a resource cartel, extracting oil and associated natural gas, facing a downward-sloping demand curve for each resource. As oil and natural gas are substitutes in consumption, Wirl demonstrates that total or partial flaring of natural gas could be optimal to support the revenues from selling oil. In particular, he shows that when oil and gas are perfect substitutes, flaring will be total during an initial period, then switching to full utilization or no flaring. When oil and gas are imperfect substitutes, the optimal marketing strategy for the monopolistic seller will be total flaring during an initial period, thereafter there will be partial flaring which diminishes over time, and then, during the last stage, to have no flaring at all. This strategy is due to the fact that the resource-rich country or cartel is a monopolistic seller in both markets.

Within our framework we have found a similar flaring strategy, but for another reason (cf. Fig. 11.1). Our framework is more in accordance with how the market for natural gas in Europe is organized. As there are few sellers and buyers of natural gas, and due to the fact that trade in natural gas requires specific capital installation, this market is more like a bilateral oligopoly, where negotiations are prevalent. When taking this specific feature into account, natural gas delivery and transfer price will be stipulated in long-term contracts. We have found that the seller may find partial flaring (which diminishes over time) to be profitable in the first part of the extraction period, then to switch to no flaring in the later part of the extraction period. Hence, the optimal flaring policy outlined by Wirl in the case of imperfect substitutes exhibits some similarities with that derived in this chapter. When a long-term relationship in gas delivery is profitable to the seller, the profitability in selling oil is the main reason for how much natural gas will be flared at each point in time, along with the size of the initial stock of oil and total gas requirement due to the gas contract. Provided that marginal willingness to pay for natural gas exceeds total marginal cost of providing the consumers with natural gas, flaring will of course represent a loss to society, but not to the seller.

ACKNOWLEDGEMENTS

The author gratefully acknowledges the constructive comments of Olav Bjerkholt, Michael Hoel and Øystein Olsen on an earlier draft.

REFERENCES

Abodunde, T. T. and Wirl, F. (1985) Optimal production of oil and gas. *Engineering Costs and Production Economics*, 9, 105–11.
Grout, P. A. (1984) Investment and wages in the absence of binding contracts: a Nash bargaining approach. *Econometrica*, 52, 449–60.
Hart, O. P. and Moore, J. (1988) Incomplete contracts and renegotiation. *Econometrica*, 56, 755–85.
Hoel, M. and Vislie, J. (1987) Bargaining, bilateral monopoly and exhaustible resources, in *Natural Gas Markets and Contracts* (eds R. Golombek, M. Hoel and J. Vislie) Elsevier Science Publishers B. V. North-Holland, Amsterdam, pp. 253–65.
Newbery, D. M. G. (1981) Oil prices, cartels, and the problem of dynamic inconsistency. *The Economic Journal*, 91, 617–46.
Pindyck, R. S. (1982) Jointly produced exhaustible resources. *Journal of Environmental Economics and Management*, 9, 291–303.
Vislie, J. (1986) Joint production and market structure: The case of oil and natural gas. *Journal of Economics (Zeitschrift für Nationalökonomie)*, 46, 163–73.
Vislie, J. (1987) Long-term bilateral contracts for natural gas, in *Natural Gas Markets and Contracts*, (eds R. Golombek, M. Hoel and J. Vislie) Elsevier Science Publishers B. V. North-Holland, Amsterdam, pp. 267–77.
Wirl, F. (1987) Joint production of substitutable, exhaustible resources, or: Is flaring gas rational? *Journal of Economics Dynamics and Control*, 11, 499–511.

Part Three

The World Oil Market and Macroeconomic Performance

12

The options for independent oil-exporting countries in the 1990s

KJELL BERGER, OLAV BJERKHOLT AND

ØYSTEIN OLSEN

12.1 INTRODUCTION

When OPEC was founded in 1960 the main aim of the organization was to secure stability in oil prices. OPEC was not able to exert a controlling influence on the world oil market until more than a decade later. It may seem paradoxical that stability in the oil price and in oil earnings has been much lower after OPEC achieved its supreme position than before, but then it was not only a question of stability, the real issue was the right to acquire the rent value of the vast low-cost OPEC oil resources.

The concept of non-OPEC oil-exporting countries is of more recent origin than OPEC itself. It gained importance as more than a residual designation in the early 1980s when it was clear that the reduced position of OPEC in the oil market (from a peak production of 31 million barrels day^{-1} (mbd) in 1979) was due not only to reduced demand but also sharply increasing non-OPEC oil exports. Among the non-OPEC oil-producing countries a group of countries has emerged for which oil production is a major industry and oil export the major source of foreign currency, as it is for the OPEC countries. Some of these countries have a long-term interest in the importance of oil as a major source of world energy, because their reserves are large relative to current production. This is particularly true for Mexico, Oman and Norway. In this regard the OPEC countries themselves constitute a rather heterogeneous group. Only some of the OPEC countries – in particular Saudi Arabia, Kuwait, Iran, Iraq, Libya and the United Arab Emirates (UAE) – have oil reserves that make it likely that oil will play a major role into the second quarter of the next century.

The history of the oil market is abundant in examples of how limited the ability to forecast the oil market has been. The future energy shortage depicted after OPEC I and the price scenarios worked out in the immediate aftermath of OPEC II are perhaps the most prominent examples. The price fall in 1986 has caused hardship in several oil-producing countries, but it has also a sobering effect on the future outlook for oil prices, perhaps to the extent of emphasizing a too pessimistic picture.

The future development of the oil market depends greatly on the extent to which the OPEC members are able to coordinate their production decisions. Countries both inside and outside OPEC will suffer large reductions in incomes by a fall in oil prices as a result of a breakdown of OPEC, as happened in 1986. The often repeated prediction that OPEC will collapse as an organization has up to now been proven wrong. OPEC has survived as an organization in spite of external pressure and extreme internal conflict. However, after the dramatic fall in oil prices in 1986, everyone has become more alert to the importance of OPEC cohesion. Some oil-producing countries outside the organization have engaged in tacit agreements with OPEC, and restricted their crude oil production in order to support OPEC's efforts to stabilize prices.

The purpose of this chapter is to analyse the options in the medium term for non-OPEC countries highly dependent upon petroleum revenues. In what follows we first outline a picture of the oil market to the year 2000 in terms of a **reference scenario** by means of a simulation model developed by the Central Bureau of Statistics of Norway (CBS). In this model, OPEC behaviour is of central importance for the projected market development. By means of a modified model framework, a **breakdown scenario** is simulated. We then look at differences between oil producers, and argue that a group of oil-producing countries, called IPEC (Independent Petroleum Exporting Countries), may have a common interest in cooperation with OPEC. In the **'cooperation' scenario** total output from the IPEC countries is reduced by 10% compared to the reference scenario. Finally, we attempt to calculate the benefits of cooperation from a Norwegian point of view.

12.2 A REFERENCE SCENARIO FOR THE OIL MARKET IN THE 1990S

12.2.1 Model framework

Over the last 15 years strong fluctuations have taken place in the international market for crude oil, with the two oil price shocks in 1973–74 and 1979–80 and the sharp price fall in the winter of 1985–86 as the most noticeable events. These price movements underline the overall uncertainty in this market, caused not only by technological and structural features of supply and demand, but also by institutional and political factors. Few – if any – econometric models have been able to capture the strong price

A reference scenario for the oil market in the 1990s

fluctuations that have actually occurred in recent years. Nevertheless, a formal model can be useful when analysing and discussing scenarios of the crude oil market, especially in a medium- and long-term perspective. In the particular, it provides an effective tool for carrying out the kind of impact calculations that we are presenting in this chapter.

Broadly speaking, one can distinguish between two classes of models available for analysing the oil market. The first consists of **optimization models**, based largely on the theory of exhaustible resources ranging back to the work of Hotelling (1931). The problems with these models as to their ability of describing actual market behaviour are well known (problems with intertemporal equilibrium, oligopoly behaviour and different solution concepts, endogenous interest rates – see e.g. Dasgupta and Heal, 1979, Hoel, 1981 and Newbery, 1981). The other class of models – **simulation models** – covers a wide group of frameworks, with generally less emphasis on theoretical consistency, stressing more operational aspects and the empirical properties of the models. In these models, special attention is naturally given to the modelling of OPEC behaviour. A common specification is the inclusion of a so-called 'reaction' function, i.e. an increasing relation between the oil price (or the rate of price change) and the capacity utilization of OPEC (see e.g. EMF, 1981). Most modellers avoid relating the relationship to any strict rational behaviour. Rather they regard it as a 'rule of thumb' for OPEC pricing decisions: increase price when the market is tight, and let it ease off when market is sluggish (Gately, 1984). Fitted to historical data, the relationship measures the 'average' price responsiveness over a specific time period. A main problem with the 'reaction' function approach is that it may not be suitable for analysing significant changes in behaviour or market conditions facing OPEC.

The **WOM model** (Lorentsen and Roland, 1985) developed in the CBS is a rather simple simulation model for the international oil market. On the demand side the model distinguishes between three regions: the USA, other OECD countries (ROECD) and less developed countries (LDCs). The demand equations in the model are of a traditional type, with demand as function of the oil price, the price competitive fuels and the income level. There are lag effects in the price responses. On the supply side, OPEC behaviour is represented by a 'reaction function', with the rate of change in the crude price increasing with utilized capacity. Capacity decisions of the OPEC countries are not explained by the model. The oil supplies from non-OPEC producers are calculated in separate submodels. As indicated above, the oil-producing countries outside OPEC are divided into IPEC and other (fringe) producers respectively. IPEC consists of Egypt, Oman, Mexico, Malaysia, Angola, Brunei, Colombia and Norway. The structure of these submodels in WOM is based on Weyant and Kline (1982) and is identical for the two groups of non-OPEC oil suppliers (to distinguish between the two is, however, essential when we come to the 'breakdown' and 'co-operation' scenarios, see below). The submodels contain some simplified dynamic mechanisms in order to capture lag effects in exploration and extraction of crude oil. The net exports of oil from centrally planned economies (CPEs) is fixed exogenously. The WOM model contains no explicit dynamic optimization mechanism, and accordingly no 'Hotelling rule' for the evolution of the oil price is obtained in the model solution.

The WOM model is used to generate a reference scenario for the oil market from 1988 (the base year) until the year 2000. As stressed above, the reference scenario for the oil market is not constructed primarily to represent a probable development. The scenario, however, is obtained within a consistent formal framework and represents in our opinion a reasonable starting-point for discussing alternative developments in the crude oil market.

12.2.2 Basic assumptions

On the supply side of the model the most critical assumption concerns the future cohesion and strategy of OPEC. In the present context we conduct the simulation with OPEC treated as a homogeneous group and with responses derived basically from historical correlations (see the discussion of the reaction function above). When calibrating the non-OPEC supply submodel, our hypothesis has been that the sharp increase in production observed over the last 10 years has come to an end, but that the current price level of $US15–20 per barrel is sufficient to support production approximately at the current level.

On the demand side assumptions have to be made concerning income growth and technological improvements in the years to come, including, for example, the development of alternative technologies to oil-using equipment. In this reference scenario, we have basically prolonged the current growth trends, i.e. we assume continued moderate income growth in the OECD countries (2.5% p.a.) and somewhat stronger group (3.5% p.a.) in LDCs. These assumptions may perhaps be said to represent 'the conventional wisdom' regarding international economic growth. The modification that debt problems may tend to slow down the growth in some LDCs is also standard.

It should be noted that due to the specification of demand functions in the model, energy substitution occurs only as a result of changes in fuel prices. With reference to a specific base year, the energy demand may, however, be out of equilibrium, and therefore substitution may actually take place even with unchanged prices. In the present model calculations, when calibrating the demand equations we have implicitly assumed that the price volatility from 1986 which brought the price down to a range of $US15–20 per barrel, basically restored a price level for crude oil where in the long run it can compete effectively with other fuels.

12.2.3 Empirical results

The simulated growth path for the oil price is shown in Fig. 12.1. The crude oil price, starting out at about $US15 per barrel in 1988, is rather flat for the first couple of years. This is mainly due to the specified OPEC reaction function. Given the modest capacity utilization in the base year, OPEC decides to keep the oil price stable by increasing production (Fig. 12.2). However, increased capacity utilization gradually motivates OPEC to loosen the price 'anchor' and combined with income growth in consuming countries this results in a gradual increase in the oil price during the 1990s. The price increase restricts demand, and accordingly OPEC production flattens out at a level of 27–28 mbd.

In Fig. 12.2 we have also shown the calculated production profile for the two groups of non-OPEC countries. For both, a rather low price responsiveness is assumed (see above). The supply from the IPEC block is then rather stable, while production from the fringe producers decreases somewhat. The latter development is due to the fact that some of the countries empty their reserves while at the same time prices are too low to motivate sufficient new exploration activities.

On the demand side, consumption of oil increases in all regions in the first 5 years due to income growth and rather low prices (Fig. 12.3). From

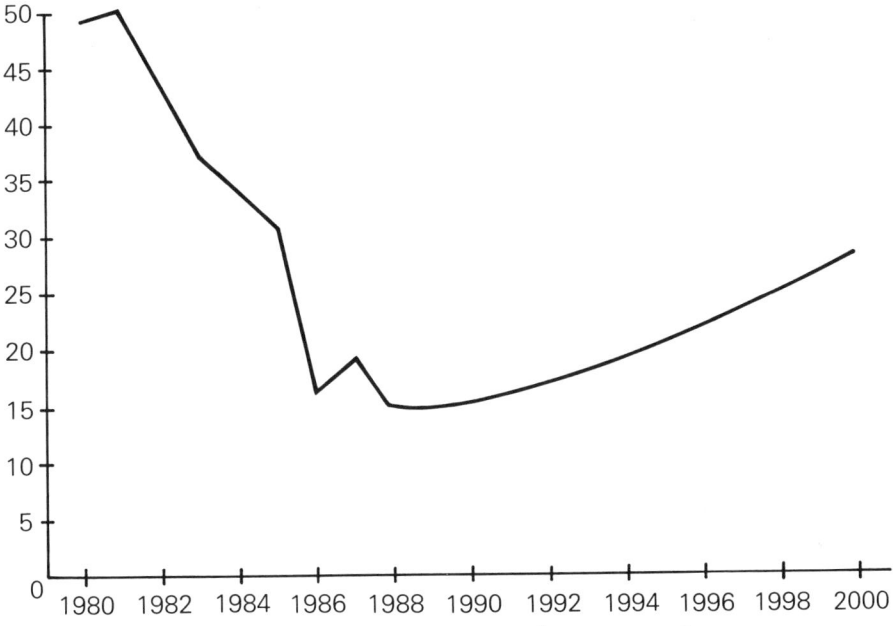

Fig. 12.1 The oil price in the reference scenario.

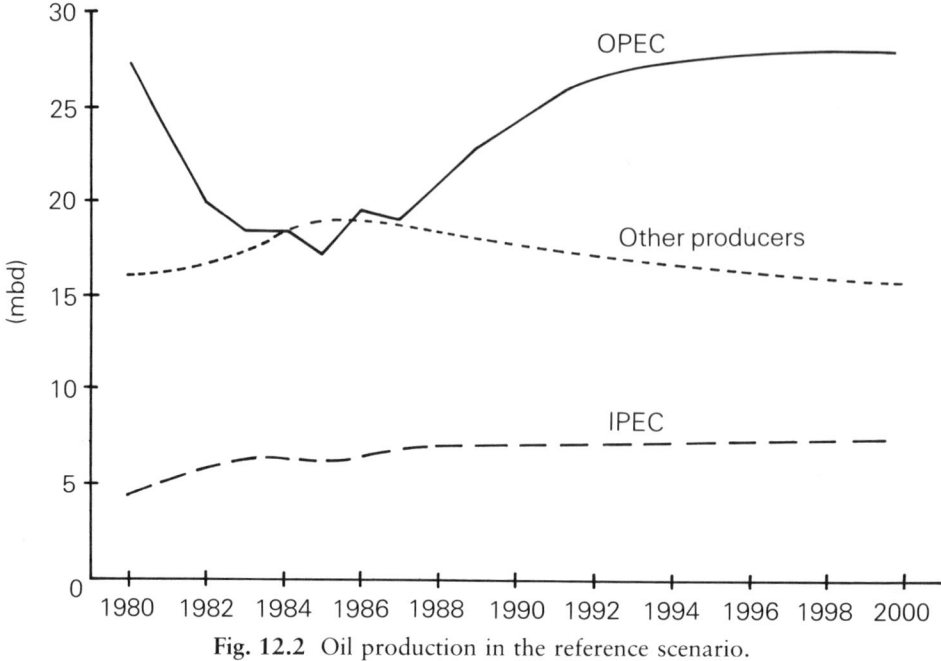

Fig. 12.2 Oil production in the reference scenario.

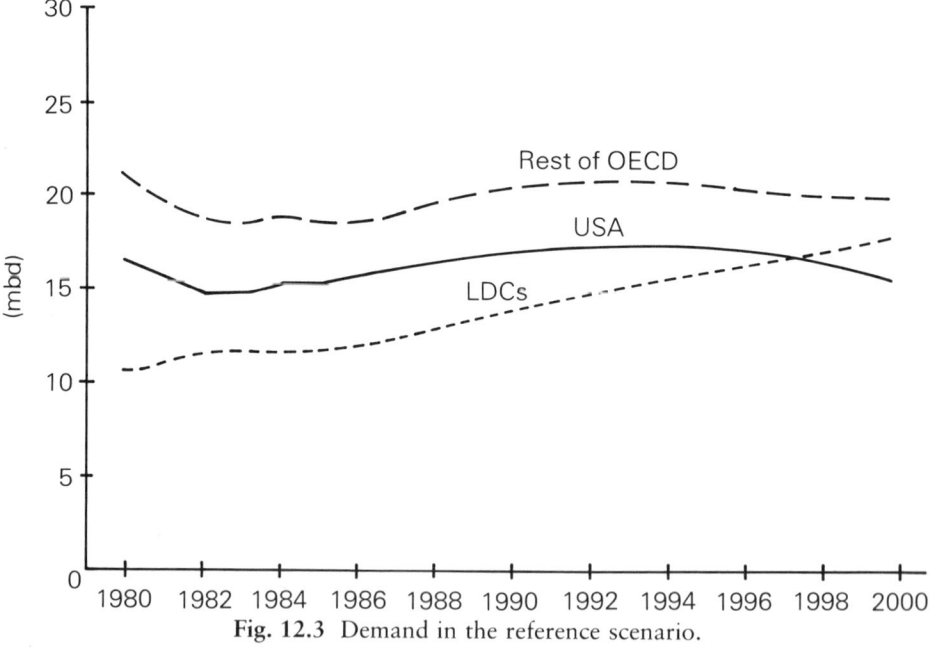

Fig. 12.3 Demand in the reference scenario.

the mid-1990s, the price increases lead to a peak in oil comsumption for the OECD area; throughout the rest of the simulation period the demand in these countries decreases at a rate of 1–1.5% per year. In the LDCs, on the other hand, due to stronger income growth and less price responsiveness in demand, consumption continues to grow towards the turn of the century.

In sum, the reference scenario simulated on the WOM model has corroborated the possibility that the crude oil price can remain at a rather low level for the coming 2–3 years. This result is based on the assumption that the present price level is sufficiently close to being an equilibrium price. The market development through the 1990s will to a great extent depend on the rate of growth in the world economy, as indicated by the calculations. In particular, one should emphasize the significance of the economic growth in the LDCs. With the assumptions made, the oil price in the reference scenario shows significant growth during the 1990s and is close to $US28 per barrel in the year 2000.

Another major uncertainty for the future development of the oil market is OPEC behaviour. Will these producers continue to co-ordinate pricing and production decisions? To what extent will the countries engage in exploration activities and increase their capacities? What will be the effects on the market if the organization collapses? As mentioned in section 12.1, these questions are also important for oil producers outside OPEC. These are problems to which we now turn.

12.3 WHAT HAPPENS IF OPEC BREAKS DOWN?

The OPEC countries, although they are all very dependent on oil incomes in their total exports, constitute a rather heterogeneous group. First, there are significant differences in income levels between the different countries. This implies different needs for foreign currency or different capacities to absorb huge incomes created by oil activities. Second, the size of the resource base varies considerably between the OPEC members, from Saudi Arabia with nearly 20% of the world's total reserves and a reserves/production (R/P) ratio of more than 100 years, to Indonesia with an R/P ratio of approximately 17 years. The resource base is a very important factor in consideration of market strategy. Countries with relatively limited reserves may prefer to cut production and increase prices to get higher profits in the short run. Countries with larger proven reserves tend to put more emphasis on the long-term outlook for the oil market.

Various tensions and political alignments play an important role when determining future market strategy. On the basis of such differences OPEC member countries are often divided into three subgroups (see e.g. Eckbo, 1976, and Berger, Bjerkholt and Olsen, 1987): the **low-income countries**, the **price hawks** and the **long timers**. In Berger, Bjerkholt and Olsen (1987) the first group comprises Ecuador, Gabon, Indonesia, Nigeria and Venezuela. The main characteristics of these countries (in addition to the fact that they are poor) are high-capacity utilization in oil activities, low R/P ratios and deficits in the current account. The price hawks consist of Algeria, Libya and Iran. For these countries political motives play an important role, and they generally tend to put priority on reaping the benefits of high prices. The last group comprises all the Gulf countries except Iran. These have generally small populations, highly oil-dependent economies and high R/P ratios. These factors make long-term market developments vital in their considerations.

The collapse of OPEC as an organization able to influence the oil policy of its member countries, has been predicted at various times since the birth of the organization in 1960, but it has survived in spite of many differences and difficulties. There is probably no other example of an international economic organization with such extreme internal conflict still being able to function. From an economic point of view, however, the interesting question is to what extent OPEC is able to control the production decisions of its members. A collapse of OPEC as an organization can hardly be expected in the near future. However, in particular during the last 5–6 years, problems of internal discipline have been frequent. One of the unsolved problems within OPEC is to work out a formula to set fair quotas on the basis of the differences in characteristics of each member country. Therefore, what should not be ruled out is the possibility that OPEC for a period will abandon its efforts on co-ordination and be unable to control the oil market.

To analyse this scenario we have utilized an oligopoly model for the international oil market where demand and supply relations are consistent with the WOM model (Berger *et al.*, 1988). The model describes a static Cournot game between a number of oil-producing countries: the thirteen OPEC countries and the individual countries which we included in the group called IPEC. In all these non-OPEC producing countries, oil plays a significant role in net exports. Furthermore, the possibilities for postponing extraction activities are limited, due either to institutional factors, low income levels, large current account deficits or capital-intensive technologies. As stated above, several of these countries have intimated that they want to take an active part in what happens in the oil market.

The remaining oil-producing countries are included in a fringe group which by assumption takes the oil price development as exogenously given. The demand functions in the WOM model and the supply structure of the fringe group yield the net demand function facing the oligopolists. Lag effects are neglected, and the demand elasticities are thus consistent with the long-term effects in demand and supply relations in WOM. For each oligopoly member, the specified cost structure is of the 'inverse-L type' marginal cost function. The capacity variables are fixed at our best estimates of today's production capacities for the various countries. The chosen values for the curvature parameters in the first years imply sharp inverse L-shaped marginal cost functions, and we thus interpret them as short-run functions. Towards the year 2000 we have changed the parameters to make the marginal cost functions more moderately inverse L-shaped, i.e. they approximate long-run functions.

To assume that the outcome in an 'OPEC breakdown scenario' can be properly analysed by a static Cournot model is surely a drastic simplification. This means that all the countries are treated symmetrically, although differences in cost structure will yield them different market shares and market power. 'Dynamic' strategic behaviour is also neglected, even though one would assume that a country like Saudi Arabia with huge reserves will be quite conscious of intertemporal aspects of market development. The Cournot solution may still represent a reasonable starting-point for discussing the effects created by a cessation of OPEC's market power. It should also be noted that with a total of 21 agents in the market, the outcome is probably not very far from a competitive equilibrium.

Using the same set of input variables as in the reference scenario calculated by WOM, the oligopoly model is simulated over the period 1988–2000. In addition we have assumed that the breakdown of OPEC will cause an increase in OPEC production capacity. In the year 2000 the increase is approximately 3 mbd. The results for crude oil production for OPEC is our assessment of what the outcome would be if OPEC breaks down. The crude oil production for OPEC is then transferred back to WOM where the 'dynamics' of the breakdown scenario are simulated. The results for the oil price and crude oil production for OPEC and IPEC are shown in Figs 12.4 and 12.5.

We have assumed that OPEC cohesion breaks down in the second half of 1989, presumably after a failed ministerial meeting. The simulation with the oligopoly model yields an estimate of OPEC production that is 5 mbd higher in 1989 than in the reference scenario. Since we assume that OPEC breaks down in the second half of 1989 the production for this year is increased by 2.5 mbd. In the 1990s the production is 2–4 mbd higher than the reference scenario.

The effect on the crude oil price calculated in WOM is quite dramatic. The price falls to $US9 per barrel in 1989, which of course implies a very low price in the second half of the year. The oil price increases to about $US10.50 per barrel in 1990 which is 30% lower than in the reference scenario. However, in 1991 the oil price jumps back to the level of the reference path as a result of a demand response to the very low price of the two preceding years. In the 1990s, still with OPEC unable to co-operate, the oil price is between 5 and 15% below the reference scenario.

The effect of the breakdown of OPEC on the production of the other crude oil producers is modest. In the year 2000 the oil production is

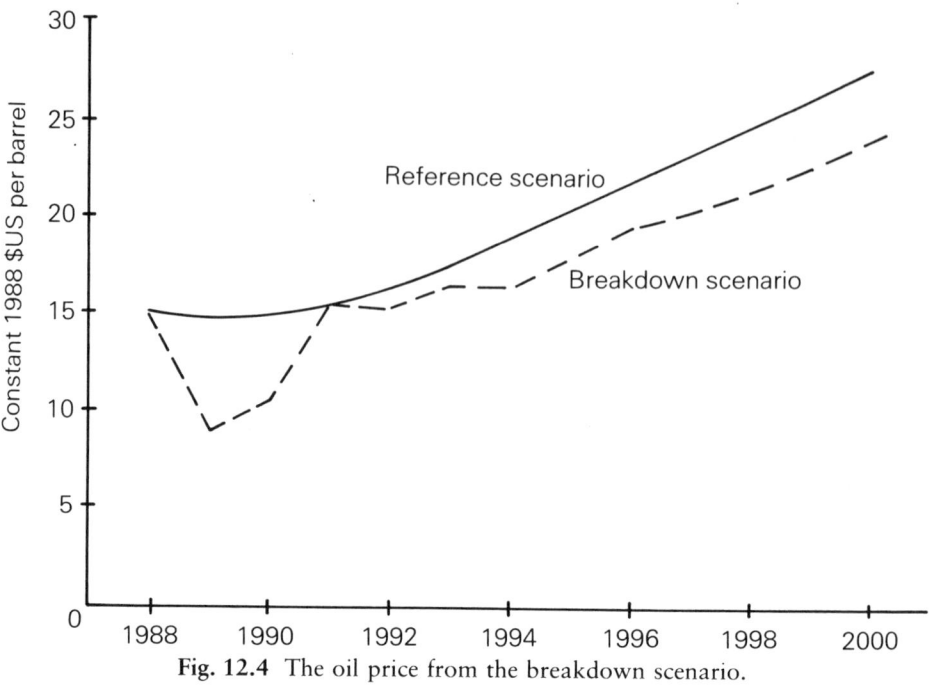

Fig. 12.4 The oil price from the breakdown scenario.

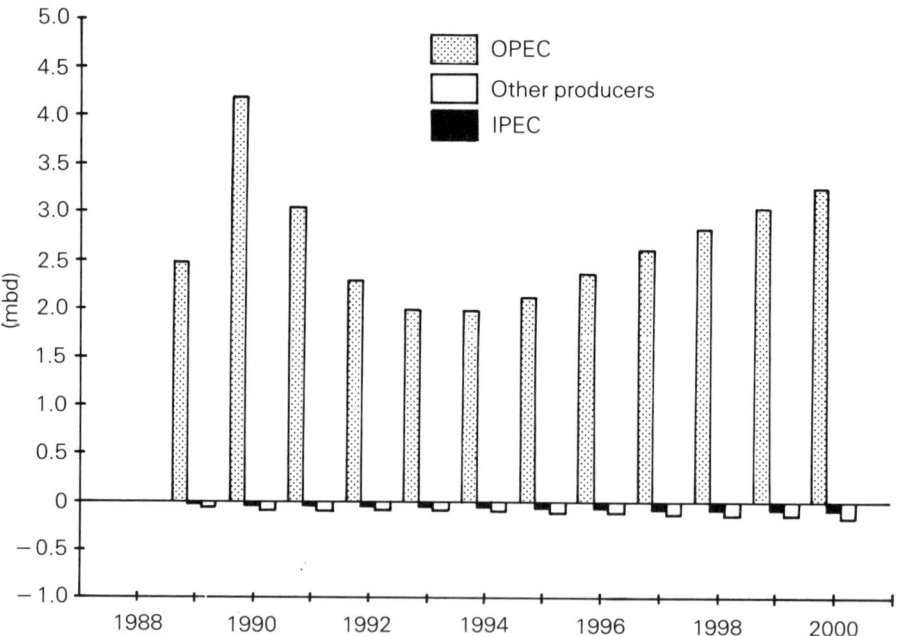

Fig. 12.5 Changes in oil production from the reference scenario.

250 000 barrels day^{-1} lower than in the reference scenario. On the demand side the largest increase is found in the OECD area due to higher price responsiveness than in the LDCs.

In our simulation of the breakdown scenario we have not explicitly assumed any renewed co-ordinating action from OPEC countries until the year 2000. This may obviously be questioned. The OPEC members will always gain if they can co-operate, and in the second half of the 1990s the total output from the OPEC countries is in the range of 30 mbd and above, while the oil price is climbing above $US20 per barrel. In these circumstances there is a significant potential for OPEC co-operation, and it is likely that the organization would sort out its problems.

Another important aspect of the breakdown scenario is the assumption that taxes on oil products are held constant during the oil price collapse of 1989–90. When the oil price plummeted in 1986 many countries increased the taxes on oil products in order to avoid a fall in the end-user price. The motives for increasing taxes were partly fiscal and partly to avoid an increase in the dependence on imported crude oil. If taxes on oil products were increased after the oil price collapse in the simulated breakdown scenario the immediate effect on the oil price would be even more dramatic. Furthermore, this would prevent the return back to the reference path after just a few years. Another aspect is that if oil product taxes are increased because of a fall in the crude oil price, the taxes are usually not decreased again even though the oil price after a period increases. The OPEC countries have to take this aspect into consideration before starting a price war.

12.4 THE REMEDY – CO-OPERATION WITH OPEC

The breakdown of OPEC and the subsequent decrease in the oil price will imply a significant reduction in total incomes from oil activities compared to the reference scenario for non-OPEC producers (estimates on the effect on total discounted incomes for Norway are presented in Table 12.2). Recognizing this, oil producers outside OPEC have a clear incentive to reach an understanding with OPEC and undertake actions that may prevent the breakdown and its attendent price crash. As mentioned above, various steps towards co-operation with OPEC have already been taken by some independent producing countries, among which are several oil producers included in IPEC. The discussions and negotiations with OPEC have taken different forms. Middle East countries like Egypt and Oman have on various occasions met and co-operated openly with OPEC. The same applies to Mexico, which has a self-imposed export quota at about 1.3 mbd. Norway has no formal agreement with OPEC, but has since February 1987 supported OPEC with an unconditional production cut of 7.5% below capacity.

Table 12.1 Crude oil production and reserves in large non-OPEC countries in 1987

	Production (mbd)	Reserves (1000 million tonnes)	R/P ratio (years)
Mexico	2.87	6.8	47.9
Norway	1.07	1.9	37.3
Oman	0.57	0.6	19.3
Egypt	0.92	0.6	12.9
Malaysia	0.51	0.4	15.3
Colombia	0.39	0.2	11.3
Angola	0.36	0.2	8.9
Brunei	0.14	0.2	27.8
Total of above	6.83	10.7	31.3
USSR	12.74	8.0	12.9
China	2.68	2.4	18.2
USA	9.91	4.2	9.0
Canada	1.91	1.0	11.3
Brazil	0.59	0.3	11.0
UK	2.61	0.7	5.5

Source: BP (1988).

During the last year and a half more concerted and co-ordinated actions towards co-operation with OPEC have been initiated by a number of IPEC countries (PIW, 1989). At a meeting in March 1988 seven IPEC countries met and agreed to cut their exports by 5%, if OPEC took similar action. Even though OPEC at this point was not ready to co-operate, a process of closer co-ordination was initiated. In November 1988 after a long period of overproduction, OPEC reached an agreement to cut production to 18.5 mbd. In the improved climate for co-operation, a number of non-OPEC producers met in January 1989. This time the group of countries included the USSR, and more informal contacts were also made with individual US producers. The representatives agreed to recommend to their governments that exports be reduced. OPEC was informed about this decision. During February and March several OPEC countries decided upon and carried out reductions in their oil supply. China, Egypt, Malaysia, Mexico, Brunei, the USSR and Oman announced a 5% production cut. Norway sustained its 7.5% below capacity production, and Angola, North Yemen and Colombia agreed to freeze production at present levels. Altogether, the present supply reductions amount to 300 000 barrels day^{-1}.

In the contacts between OPEC and the IPEC countries the common objective has been to achieve the $US18 per barrel price target. So far, the actions may be said to have been a success, even though specific interruptions of supply and surprisingly strong demand growth may have made the task easy. An important question is, however, whether co-operative efforts

The remedy – cooperation with OPEC

are sufficient to stabilize the oil price for a longer period, say towards the year 2000.

To study this problem analytically we again turn to the WOM model, and use this to analyse the equilibrium in the oil market caused by a self-imposed reduction in the supply from non-OPEC countries. The production cuts are assumed to be prolonged and extended from present agreements (remember also that the base year is 1988). More specifically we assume that the countries included in the IPEC group acting independently, but with some sort of understanding with OPEC, cut their total supply by 10% compared to the reference scenario. We also assume that similar actions are taken by the USSR and China. Altogether non-OPEC supply is reduced by a little more than 900 000 barrels day^{-1} yearly. Technically within WOM, the supply of IPEC is reduced exogenously. The reductions are assumed to be initiated in 1989 and last until 1996, when the oil price becomes sufficiently strong so that there is little incentive to carry further supply reductions.

An important question is how the reductions in IPEC supply affect OPEC behaviour. As stated above, OPEC has shown a clear interest in discussing these matters with IPEC, and regard their supply reductions as conclusive evidence of their efforts to stabilize prices. However, if the general market

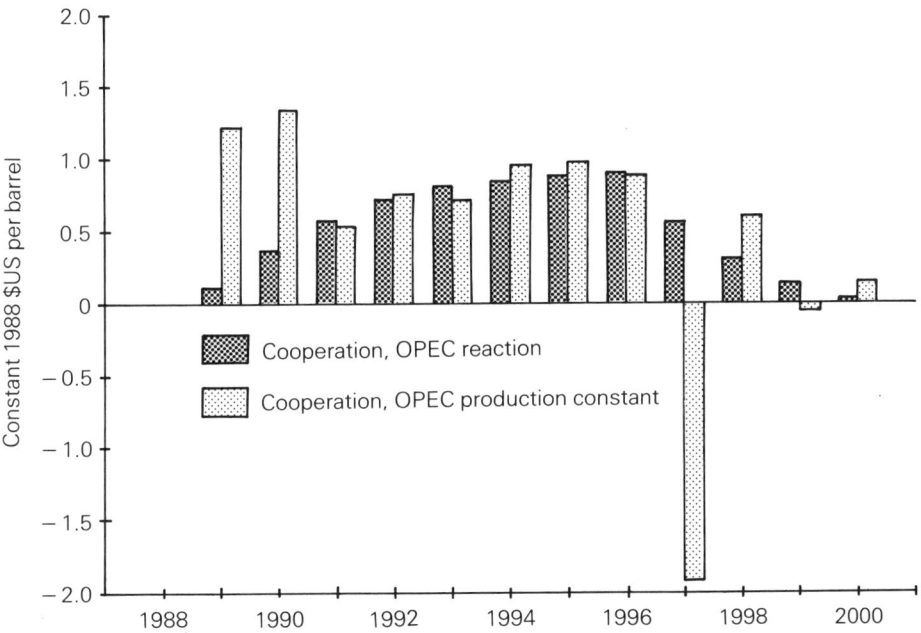

Fig. 12.6 Changes in the oil price from the reference scenario.

conditions continue to strengthen, there is certainly no guarantee that the situation will not be exploited by OPEC, seeking to increase its market share. With this background we have simulated the WOM model with 'co-operation between IPEC and OPEC' with two different assumptions of supply responses from OPEC. In the first simulation, OPEC production is simply kept constant from the reference scenario. In an alternative simulation, OPEC responds according to the reaction function specified in WOM.

The resulting impact on the oil price development is shown in Fig. 12.6. Since total supply to the market is reduced, the oil price exceeds the price path in the reference scenario, as long as co-operation takes place. Initially, the price reaches the highest levels when OPEC production is kept unchanged. When OPEC uses the tightened market to increase capacity utilization, this tend to dampen the price rise. On the other hand, in the latter scenario, a fall in prices when co-operation ceases is avoided; when supply suddenly increases in IPEC and in CPEs, OPEC supports the high prices by immediately reducing its production (Fig. 12.7). With no supply reaction by OPEC, there is a temporary fall in oil prices when new IPEC supply is brought into the market. However, in both simulations continued growth in demand brings the price close to $US28 per barrel around the year 2000.

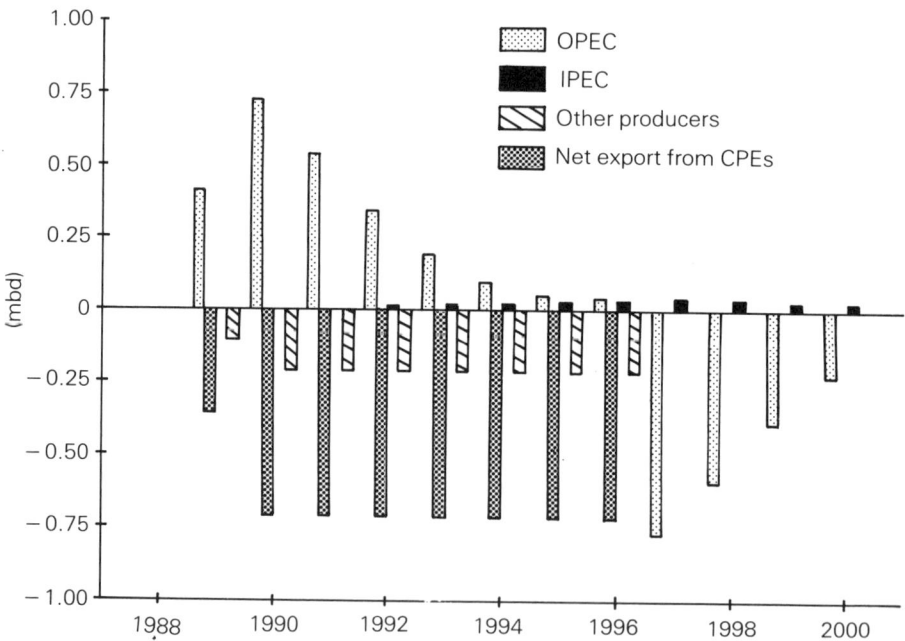

Fig. 12.7 Changes in oil production in the co-operation scenario, with OPEC reaction function.

12.5 IMPACTS ON NORWAY'S INCOME FROM OIL PRODUCTION

The four price paths outlined earlier in this section obviously have very different impacts on the Norwegian income from oil activities. We have assumed that without co-operation with OPEC Norwegian oil production will grow to about 2 mbd in 1994, and then be reduced somewhat towards the turn of the century when it is estimated at 1.4 mbd. We assume that the same production path is sustained by Norway in the breakdown scenario, presupposing that profitability on the Norwegian continental shelf under the present tax regime is sufficiently favourable even in this case.

In the co-operation scenario the Norwegian production is reduced by 10% from the second half of 1989 to beyond 1996. During the rest of the century production is assumed to reach the same level as in the reference scenario which implies a somewhat lower total production over the simulation period.

Based on the four price scenarios and the corresponding assumptions regarding oil production, we have calculated the net present value (NPV) of Norwegian oil income in the four cases with a 5% interest rate. As shown in Table 12.2, not surprisingly the NPV is highest in the reference scenario, where we have obtained an estimate of NPV of $US104.3 bn. Moreover, the calculations show that there is only a modest reduction in NPV in the two co-operation scenarios since a part of the reduced volume of production is compensated with higher price. The most striking feature is, however, the large decrease in incomes in the breakdown scenario, where NPV is reduced, by $US13 bn compared to the reference scenario.

Table 12.2 also shows the NPV of gas production. Norway, like several other oil-rich countries, is also a large gas producer. With gas reserves that exceed the known oil reserves, the future development of gas prices is very important. We have also computed the NPV of the expected Norwegian gas exports, and presented the figures in Table 12.2. The gas price is assumed on the basis of historical data to be 65% of the previous year's oil price. Norwegian gas export volumes will undergo a slump to 22 bcm in the early

Table 12.2 Net present value of oil and gas production in the four scenarios (bn 1988 $US)

	Oil	Gas	Total
Reference scenario	104.3	23.5	127.8
Co-operation scenario, constant OPEC production	100.6	24.0	124.6
Co-operation scenario, reduced OPEC production	100.4	24.0	124.4
Breakdown scenario	91.2	21.0	112.2

1990s before the Troll/Sleipner contract comes into full effect. The NPV of gas exports is highest in the co-operation scenario because of the higher price in the early years and no change in volumes. There is only a small reduction in the reference scenario, while the NPV in the breakdown scenario is $US3 bn lower.

For the total value of oil and gas production our estimates indicate that total NPV is $US3.2 bn lower in the co-operation scenario than in the reference scenario, while the NPV through co-operation is more than $US15 bn higher than in the breakdown case. Thus, the calculations demonstrate that it may be very important for Norway to avoid the breakdown scenario. The 10% reduction of the eight countries alone does not significantly influence the oil price, but if this effort is sufficient to keep OPEC together it is an important step towards stable higher prices. The reduction in NPV in the co-operation scenario can be seen as a small insurance premium well worth paying for a better chance to avoid the breakdown scenario.

12.6 SUPPORTING OPEC: WHAT ARE THE RULES OF THE GAME?

The simulations presented above serve to illustrate that the future stability in the oil market is highly important for non-OPEC producing countries. What makes it so difficult to choose strategy for these countries is of course that considerable uncertainty exists with respect to the future strategy and market power of OPEC. Producers outside the organization will reap the highest benefits if market development is not dependent on their active co-operation so that they can continue as free riders. There may be some signs in the market development following the price fall in 1985–86 that may be used as an argument for such a choice. After all, OPEC seems to have succeeded well so far in stabilizing the market with rather moderate support from other countries. On the other hand, there is obviously a possibility that due to tensions and differing goals and interests within the organization, the agreement of market regulation is violated, with a sharp decrease in the oil price as the likely result. Aversion against this outcome will be the motive for producers outside the organization to enter some sort of co-operation with OPEC. This may be interpreted as an insurance against a breakdown of OPEC. As we saw above, the 'risk premium' in terms of income loss, with Norway as an example, may be relatively small.

The key question is thus whether OPEC will be able to keep together and retain the ability of co-ordinating action. From historical experience there seems to be little reason for expecting a breakdown in the immediate future. One can for instance point to the strong recovery of the organization when it managed to pull itself together in the summer of 1986 and stabilize prices

Supporting OPEC: what are the rule of the game?

temporarily, and the result of the November 1988 meeting. It therefore seems reasonable to conclude that OPEC will manage to keep together in spite of both internal conflicts and differing views and interests with regard to pricing policy.

Still, with respect to the future of OPEC, the collapse of the organization as an effective market regulator cannot be ruled out. A long-contested issue is whether OPEC is actually a cartel exerting monopoly control of the market. OPEC can hardly be described in terms which explains its actions since 1973, not to mention the earlier period, as derivations from underlying rules of cartel behaviour. The aim of a cartel is to extract the maximum benefit for its members using the control over the members' production. In a market with various disturbances and uncertainty on the demand side this aim requires external flexibility and internal discipline. The external flexibility is necessary to exploit the market conditions at all times. This, in turn, presupposes internal discipline. The executives of the cartel must have the authority to act on behalf of all and every one of the members.

The president of OPEC has seldom been in this situation. Only for a very short period of its history has OPEC adhered to the cartel model of distributing binding member quotas. The internal discipline has never been strict – after all the members are not merely profit-making concerns but national governments. Internal conflicts and disagreements have thus been frequent. Furthermore, the outward side has not been nearly as flexible and foresighted as expected of an efficient cartel. Too often exogenous events have run their own course with little correction from the organization. Looking back at the history of OPEC the ratchet effects are obvious, especially with reference to the events in 1979–80. Saudi Arabia, in particular, expressed at the time a clear conviction that prices had been carried too far above a reasonable level, but for several years had to acquiesce to the policy.

From history we have thus learned that when OPEC is under pressure from internal or external forces, one may expect the actions of the organization to be sluggish and – in view of the low short-run price elasticity – allow wide swings in prices. At this point there is scope for contributions and actions from non-OPEC producers, influencing the cohesion of OPEC. Oil-producing countries outside OPEC have an obvious interest in stable and relatively high prices. Recently, the efforts in the direction of co-ordination of IPEC supplies and explicit steps towards co-operation with OPEC have been intensified. OPEC has also clearly indicated an interest in co-operating with non-OPEC countries.

The contacts between OPEC and non-OPEC countries have been of different kinds, as mentioned earlier. The Norwegian government has announced reductions relative to full capacity production for a limited period ahead. It has furthermore made this agreement conditional on OPEC being able to keep control over its own production. Other countries within IPEC have practised more direct contacts with OPEC countries.

An important aspect of OPEC–IPEC relations is the non-OPEC countries' room for manœuvre in adjusting production. This depends upon institutional conditions with regard to government control over current production and capacity utilization as well as technical and cost conditions. It is worth noting that although Norwegian offshore production remained below capacity, it has increased quite dramatically, especially during the last 2–3 years. It will increase further according to existing investment plans. Other non-OPEC producers may be much more constrained in this respect. Considering the IPEC countries as a group should theoretically open up possibilities of greater room for manœuvre. However, it is not at all obvious that a joint non-OPEC stance towards OPEC would achieve more than individual arrangements. A 'united' IPEC could be considered as a threat, while the present shuttle diplomacy practised by OPEC *vis-à-vis* the non-OPEC countries may be seen as a support to the central role of OPEC even when its percentage share is shrinking.

REFERENCES

Berger, K., Bjerkholt, O. and Olsen, Ø. (1987) What are the options for non-OPEC producing countries? Discussion Paper no. 26, Central Bureau of Statistics of Norway.

Berger, K., Hoel, M., Holden, S. and Olsen, Ø. (1988) The oil market as an oligopoly, Discussion Paper no. 32, Central Bureau of Statistics of Norway.

Dasgupta, P. S. and Heal, G. M. (1979) *Economic Theory of Exhaustible Resources*, Cambridge University Press.

Eckbo, P. L. (1976) *The Future of World Oil*, Ballinger, Cambridge, Mass.

EMF (1981) *Energy Modelling Forum, World Oil, EMF 6*, Summary Report, Stanford University, December.

Gately, D. (1984) A ten-year retrospective: OPEC and the world oil market. *Journal of Economic Literature*, 22, 1100–14.

Hoel, M. (1981) Resource extraction by a monopolist with influence over the rate of return on non-resource assets. *International Economic Review*, 22, 147–57.

Hotelling, H. (1931) The economics of exhaustible resources. *Journal of Political Economy*, 39, 137–75.

Lorentsen, L. and Roland, K. (1985) Modelling the crude oil market. Oil prices in the long term, In *Macroeconomic Prospects for a Small Oil Exporting Country* (eds O. Bjerkholt and E. Offerdal), Martinus Nijhoff, Dordrecht.

Newbery, D. (1981) Oil prices cartels and the problem of dynamic inconsistency. *Economic Journal*, 91, 617–46.

PIW (1989) *Petroleum Intelligence Weekly*, 27 March.

Weyant, J. P. and Kline, D. M. (1982) OPEC and the oil glut: outlook for oil export revenues during the 1980s and 1990s. *OPEC Review*, 6(4), 333–64.

13

Business cycles and oil price fluctuations: some evidence for six OECD countries

KNUT ANTON MORK, HANS TERJE MYSEN

AND ØYSTEIN OLSEN

13.1 INTRODUCTION

The disruptions in the world oil market during the 1970s and the subsequent business cycle fluctuations set off a round of research on the macroeconomic effects of such price shocks. Theoretical explanations for the link between changes in oil prices and fluctuations in the overall economic activity level were presented (see e.g. Phelps, 1978), and a number of simulation studies were carried out (Pierce and Enzler, 1974; Gordon, 1975; Mork and Hall, 1980). Empirical investigations of the relationship have also been undertaken by Darby (1982), Hamilton (1983), Burbridge and Harrison (1984) and Gisser and Goodwin (1986). For an oil-importing country both theory and data point to a negative correlation between increases in petroleum prices and the overall activity level. According to Hamilton (1983), there were thus clear tendencies of stagnation in the US economy occurring six months to one year after the two oil price rises in the 1970s, as well as earlier in the postwar period. Similar impacts from the oil price shocks are estimated for other countries (Burbridge and Harrison, 1984).

During the winter of 1986, crude oil prices fell to an extent that rivalled the increases of the 1970s. The findings in the literature mentioned above predicted that such a shift downward in prices would stimulate economic activity in oil-importing economies. However, so far few signs of such positive effects on GDP have been observed in the OECD area. This experience raises two questions: (1) whether the correlations observed by Hamilton and others were spurious, in other words whether oil prices affect business cycles at all, and (2), if there is an effect, whether it is asymmetric. The asymmetry hypothesis, which says that the growth stimulus from an oil price decrease will not match the positive impacts triggered by an increase in prices, is implied by Hamilton (1988) and has been tested by Loungani (1986) and Davis (1987). However, to some extent these studies had little power, because, at the time they were carried out, no substantial oil price declines had yet taken place. The turn-round in the oil market in 1986 thus provides us with a unique opportunity of examining the issue of asymmetry.

Using data through the second quarter of (1988:2) 1988 Mork (1988) finds significant evidence of asymmetric effects from oil price fluctuations in GDP growth in the USA. Mork obtains his results within the context of a vector autoregressive (VAR) model similar to Sims (1980) and Hamilton (1983). The present chapter extends this type of approach to include Canada, West Germany, Japan, the UK and Norway in addition to the USA. A main objective of the analysis is to test for asymmetric effects of price changes in each country. By an inter-country comparison one may be able to focus, for example, on how the macroeconomic effects of price shocks depend on the relative dependence of oil imports in the economy. Two countries – Norway and the UK – have experienced significant adjustments in their industrial structure, as they have moved from being oil importers to significant net exporters of crude oil. Testing the correlation between oil prices and GDP for these countries may thus shed further light on the question of asymmetric effects of price fluctuations.

13.2 OIL PRICE FLUCTUATIONS AND MACROECONOMIC PERFORMANCE: A BRIEF SYNOPSIS OF THE THEORY

From economic theory a number of channels can be identified through which oil price changes may affect the economy's overall activity level. Obviously, a detailed discussion of the economic consequences of changes in oil prices will depend on the assumed 'model' for the functioning of the economy. In particular, the **time horizon** of the analysis will be decisive for the extent at which the economy is able to adapt to changes in relative prices. For instance, in the very short run an upward shift in the oil price will have immediate effects on the trade balance. In the medium term, such an event will trigger some sort of adjustment in economic behaviour, either autonomously by private agents or as a result of policy measures taken by public authorities. By adjustments to new relative prices through substitution responses economic agents can reduce the loss of income caused by a worsening of the terms of trade. Expectations of the future development in oil prices are important for these responses; if a price change is believed to be 'permanent', actions to tighten a gap and avoid imbalances may be taken immediately.

In this chapter, the focus is on the short- to medium-term effects of oil price changes, i.e. on business cycles. This means that we ignore long-term growth effects via capital accumulation and technical change.[1]

The main arguments in the studies on the effects of changes in oil prices on business cycles are as follows: in a one-sector model of an oil-importing country, oil can be treated as a 'third' factor in an aggregate production function of the economy (capital is fixed in the short run). When the price of oil increases, this motivates substitution away from energy and material inputs (which also will have higher prices as a result of the oil price shift) and to a reduction in the supply of products. Unless labour and energy are very close substitutes, the increased energy costs will also imply a negative shift in the marginal product of labour. These supply-side effects occur whether or not wages are flexible. In an equilibrium model, the effect on actual employment depends on the elasticity of labour supply. Wage rigidity may also cause a fraction of the labour force to be 'involuntarily' unemployed.

Furthermore, oil price shocks also affect the economy through aggregate demand effects. Unless other prices move sufficiently to offset the effect of increased energy prices, oil price fluctuations affect the overall price level, thus producing a real balance effect. This would add to the negative correlation between oil prices and the economic activity, either directly affecting consumption demand or indirectly limiting activity though the money market.

A third explanation of the postulated negative correlation points to the transfers of income between countries that take place when oil prices increase. Oil producers have their incomes increased while consuming countries suffer from a deterioration of terms of trade. If oil producers have smaller propensities to consume than oil-importing economies, one may see a general contraction in international trade and aggregate demand. The same argument can also be augmented to include domestic income transfers if it can be assumed that domestic oil producers have relatively low propensities to spend (Horwich and Weimer, 1984).

13.3 ASYMMETRY IN PRICE RESPONSES

So far, we have referred to the events in the 1970s and the more sluggish growth in oil-importing countries resulting from increases in oil prices. At the outset, one would perhaps think that all the effects mentioned above are symmetrical, in the sense that oil price declines should move real output by the same force as price increases, only in the opposite direction. Moreover,

---

[1] For this reason, the constancy of the coefficients in our model should be viewed as an approximation. The validity of this approximation may be questioned when we use data stretching over a time period of several decades.

the effects following a price change should go in opposite directions for oil importers and exporters respectively. However, as mentioned in section 13.1 so far there is little sign of an upswing in economic activity in oil-importing countries after the shift in the oil market in 1985–86. The postulated theoretical explanations arising from the events in the 1970s therefore need to be supplemented.

Clearly, a one-sector model for the economy is too restricted a framework for discussing the effects on the economy of oil price changes. Various sectors have different energy intensities, and will therefore be affected differently by price changes. Moreover, some countries have important domestic oil industries. The heterogeneity between sectors and problems created by reallocation of factors of production as an explanation of *asymmetry* in oil price responses has been discussed recently by Hamilton (1988), Davis (1987) and Mork (1988). The essential argument is that in the short run, a sudden and significant change in relative prices may create structural imbalances in the economy and produce a negative drag on the activity level whatever is the direction of the price change. Parts of the labour force and capital stock may become unemployed, and thus the aggregate output of the economy is reduced; at least for a period. For an oil importer, these problems of idle capacities will strengthen the negative effects of a price increase discussed above. However, when the price falls, there will be forces working in different directions: terms of trade are improved, but the positive stimulus from this may, at least in the short run, be offset by frictions and costs of reallocation between sectors. The net effect on GDP then becomes ambiguous.[2]

An example of reallocation problems triggered by an oil price shock may be the development in the USA after prices went down in the early and mid-1980s. As a result, the US oil industry had to cut back on activity and employment in order to survive, as did state and local governments in oil-producing areas. At the same time, the struggling airline industry was given an important relief from high fuel costs, and the automobile industry saw the prospect of a return to a trend of more and larger cars. However, to have capital and labour reallocated between sectors takes time and involves costs of adjustment. When relative prices move gradually, this may occur without too much disturbance. But when the economy is hit by dramatic price changes, bottlenecks of various kinds may impede fast adjustment. For example, structural unemployment may rise, as capital investments by and large are irreversible and labour skills are industry specific. Such imbalances will, at least temporarily, reduce the aggregate output of the economy.

For an oil exporting country, one would expect the effects on the

---

[2]As mentioned above, in the longer run an economy should be able to benefit from reallocation of resources.

economy from a price increase to be the opposite of those of an oil-importing country. Initially, the net exporter experiences an improvement in the terms of trade. If the increased incomes are absorbed in the economy, either directly affecting private behaviour or via a more expansionary economic policy, this creates a positive stimulus on the overall activity level. Calculations discussing the effects of energy price changes on the Norwegian economy (a significant net exporter of oil) are presented in Longva, Olsen and Strøm (1988).

Due to problems of reallocation there may be tendencies towards asymmetric effects between oil importers and oil exporters. For an oil-exporting country, various kinds of frictions and misallocation of resources may dampen the positive stimulus on the activity level stemming from an increase in the oil price. These effects are closely related to the phenomenon discussed in the literature under the name of 'Dutch disease', i.e. problems that may occur in an economy that has become dependent on income from petroleum activities (see e.g. van Wijnbergen, 1985). A recent example of a country that may be said to be struck by 'Dutch disease' is Norway. For this country, as for the UK – the other large North Sea producer – the two oil embargoes in the 1970s created a big boom both in the offshore industry and in industries delivering goods and services to the petroleum sector. As a result of absorption of the increased incomes in the economy there was a strong increase in the demand for non-tradables and goods from protected industries and a contraction in non-oil tradables. When prices turn around this development may be very difficult to reverse. The industrial structure, capital stocks and the labour market are adjusted to high energy price, and a restructuring of the economy to the new situation may take time and be costly to society. The impacts on the Norwegian economy of a significant fall in oil prices are discussed in Berger *et al.* (1987).

Another aspect that may be important in explaining asymmetric effects of price changes on the economy (and which may be part of the 'story' discussed above) is the economic policy that accompanies a sudden change in oil prices. Following the dramatic increases in oil prices during the 1970s there was persistent inflation in many OECD countries. As a consequence, economic policy became more restrictive, and many economies passed into stagflation. Even if some adjustments in government policy were certainly necessary in oil-importing countries, a general fear of inflation and lack of co-ordinating efforts in pursuing more expansionary policies probably played significant roles on their own in the years after the oil price shocks. There are good reasons to believe that economic policy is asymmetric in this respect, so that when oil prices decline this is not met by any corresponding expansionary policy in oil-importing countries.

Related to the question of how economic policy is influenced by an oil price shock is the aspect of uncertainty. Strong fluctuations in oil prices may imply increased perception of uncertainty. This may in itself have negative consequences on activity levels both in private business and through a more restrictive economic policy. The experiences following from the price shocks during the last two decades will probably influence expectations for a long time to come.

13.4 MODEL, METHODOLOGY AND DATA

From a theorectical point of view, it could be argued that the most preferable procedure for analysing the effects on business cycles of oil price shocks is to utilize a complete, structural model of how the economy works. However, the literature does not single out one particular model for empirical testing, and the task of including all relevant elements and mechanisms in one framework is insuperable. Instead, we have chosen to specify a reduced-form model, and to use empirical formulations that can provide useful information under a wide variety of circumstances. This was the philosophy underlying Hamilton's (1983) study of the oil price–GNP growth relationship in the US economy, using Sim's (1980) VAR model. Hamilton estimated a single reduced-form equation for GDP growth and applied univariate causality tests for examining the significance of oil price fluctuations. In the same tradition, Burbridge and Harrison (1984) estimated a complete VAR model, i.e. also explaining the change in the nominal price level, while the analysis of Mork (1988) was based on a single (VAR) GDP–growth equation. Both the latter studies focused specifically on the oil price as an explanatory variable.

In this chapter, we follow the approach of Hamilton and Mork and specify a reduced-form regression model for GDP growth. Since our allowance for asymmetric responses introduces a non-linearity in the model, it cannot be inverted by standard methods even if we had estimated reduced-form equations for all of the variables. Thus, in this sense, our model is not a true VAR model nor even an equation from one; however, we do interpret our equation as a reduced-form model of GDP/GNP fluctuations. Regarding the effects of the oil price on GDP, we carry out univariate tests as well as tests based on a more fully specified model. However, our model needs to be modified relative to that of Hamilton (1983) in order to accommodate our testing of asymmetry effects. The simple correlations are interesting because these estimates can be interpreted as the 'total' effects of oil price changes, after policy and other domestic or international responses to the oil price change have affected real growth indirectly. On the other hand, in the multiple regressions the oil price coefficient is indicative of the 'partial' effects of oil price changes. It is quite possible that the partial effects are negligible even if the simple correlations are non-zero. This could happen, for example, if oil price fluctuations have no real aggregate effects by themselves, but give rise to anti-inflationary policy measures with real economic impacts. Another example would be if a country does not experience real effects in a direct sense, but is significantly affected by trade with other countries. It should be emphasized that the definition of 'total' and 'partial' effects are ambiguous, as they obviously depend on the specific model utilized. In Longva, Olsen and Strøm (1988) the concepts of total and partial energy price elasticities are defined and discussed within a disaggregated general equilibrium model, and estimates of energy price effects for the Norwegian economy are presented.

Our basic model is specified as follows. The data used are quarterly, and the variable on the left-hand side of the VAR equation is, in each case, the country's real GDP growth rate. On the right-hand side we always include a fourth-quarter distributed lag of real GDP growth. In addition, in each model version we specify four lags of the appropriate oil price variable, to be discussed in some detail below. For the univariate models this completes the variable list. The multivariate VAR equations include the following additional variables[3]:

1. a short-term interest rate variable;
2. the rate of change in real wages;
3. the unemployment rate;
4. the inflation rate;
5. the overall index of industrial production in the OECD area.

The latter variable was included by Burbridge and Harrison as a proxy for the interdependence between countries via foreign trade. Except for the unemployment rate and the interest rate all variables were included in the regressions with their yearly growth rates, calculated as 400 times the log change of the respective quarterly figures.

The main data source for the present study has been the OECD main economic indicators (MEI). For Norway, we have utilized GDP figures from the quarterly National Accounts, and information of unemployment is collected from a specific survey (AKU). Moreover, for this country no short-term interest rate was listed in MEI, and an interest rate on long-term bonds was then substituted for this variable.[4] Both unemployment rates and GDP figures for all the countries were adjusted for seasonal variations. The starting dates for the available data series vary from series to series and from country to country. We decided to use the same estimation period for all countries which, considering the lags and first-difference specification, limited our estimation period to 1967:2–1988:2. More detailed information on the date is given in the Appendix.

---

[3] In some preliminary runs we also included the growth rate in import prices as an independent variable. However, to some extent international interactions were already represented by the index for OECD. In addition, when we experienced insignificant coefficients for the import price variable, we decided to exclude this variable from the final version.
[4] Note that interest rates in Norway were subject to regulation for a large part of the observation period.

13.4.1. Oil price variables

The construction of a relevant oil price variable is very important in this kind of analysis. In our view, some of the literature cited earlier has not given this enough attention. Different price concepts for oil exist; some are relevant for consumers, others for producers, and there are also differences between countries due to fluctuations in currencies, taxes and price controls. In Burbridge and Harrison (1984), the dollar world price of crude oil was used in the regressions for all countries. The price may be a good indicator of world markets disturbances, but has significant weaknesses as a measure of domestic costs and revenues, if taxes, subsidies, price controls or exchange rate fluctuations put wedges between the dollar crude price and the price paid or received by domestic consumers and producers. The price control schemes in the USA and Canada, the high and varying taxes on petroleum products in Europe, and the violent fluctuations in exchange rates since 1972 are all important examples of such wedges.[5] Against this background, we considered the choice of oil price variables carefully for each country. The choices were based partly on a priori arguments and partly on empirical correlations undertaken. Obviously, this procedure biases our results somewhat in the direction of finding such a correlation.

For the USA the alternatives were the world price of crude oil, measured as the spot price of 'Arabian light',[6] the US producer price index (PPI) for crude oil, the PPI for petroleum products, and the PPI for crude oil corrected for price controls as constructed by Mork (1988). The latter index represents a chaining of the PPI for crude oil with the refiner acquisition cost (RAC)(composite for imported and domestic oil), for which data have been available since the early 1970s. Following Mork (1988), we decided that the latter price index is preferable to the unmodified PPI. Furthermore, we considered the world market crude price to be unsuitable, since the USA had been sheltered from the world market by price controls during the 1970s. On a theoretical basis, the PPI for products should carry additional information as a cost indicator for oil-consuming industries and households. Empirically, however, there is a very close correlation between product prices and the combined PPI/RAC index. As a rather arbitrary choice between these two, we decided to use Mork's modified producer price series in our regressions for the USA.

For Canada, West Germany and Japan the choices were limited to the world oil price (in US dollars or converted to local currencies). For West Germany and Japan the logical choice on a priori grounds was the PPI for petroleum products, since these countries import all their required oil. The

---

[5]For a study of the development of real prices of crude oil and petroleum products in OECD countries, see Huntington (1984).
[6]Until the fourth quarter 1978, official sales prices ('posted prices') are used. Thereafter, spot prices are utilized as a better indicator for the market value of crude oil.

situation for Canada is quite different, since this country has a significant domestic oil industry. In the same way as for the USA, one should then expect the PPI price for products to represent prices paid by consuming agents in the market, and the world price of crude to be the price received by the producers. However, Canadian price policies may have caused considerable deviations between the world market price and the prices actually received by domestic producers. This may explain why we obtained a very weak correlation between changes in the world crude price and economic activity for Canada. We therefore decided to include the PPI price for products for Canada as well.

The UK and Norway have undergone dramatic changes in their economic structure during the observation period, as they have switched from being net importers of oil to being significant exporters. Our model specification for these countries should allow these transitions to be reflected in the estimation results. Since oil producers in the UK and Norway have been free to sell their oil at the world price, our data series for the world price of crude, converted into local currencies, should be a good indicator of the marginal revenue to crude oil producers. As the North Sea production becomes substantial, one should expect a positive correlation between oil price fluctuations and changes in GDP. As for other countries, the domestic PPI for products should reflect the marginal price of oil in oil-consuming activities.

After having experimented with quite a large number of different specifications for the UK and Norway, we settled on a model with three oil price variables included. In addition to the PPI, we also introduced two variables for the world price of crude in local currency, distinguished by the fact that they take on zero values for the period before/after each country became a net exporter of crude oil.

13.4.2 Measures of asymmetry in price responses

So far, we have bypassed the question of how to specify the oil price variable in the regression models so that we are able to test for asymmetric effects of price fluctuations on economic growth. In the literature, various methods have been utilized. Loungani (1986) and Davis (1987) faced this problem by adding the squares of the lag changes in the price of oil in the equation to be estimated. Another possible solution is to undertake tests for structural changes between periods with mostly price increases and periods with falling prices. In this chapter we follow Mork (1988) in specifying price increases and price decreases as separate variables and estimating separate coefficients independently of each other.[7]

---

[7]It may be noted that this distinction implies that the regression model is non-linear. This furthermore means that when specified as a VAR model it cannot be inverted by standard methods.

More precisely, the above-mentioned specification was used in the econometric models for the USA, Canada, West Germany and Japan. For the two net exporters of oil in our sample, the UK and Norway, no distinction between variables for 'price increase' and 'price decreases' was made. The reason is that for these countries we had already used up degrees of freedom by distinguishing between producer and consumer prices respectively as discussed below. Clearly, in a sufficiently detailed structural model one should be able to identify asymmetric effects of different kinds in the economy, in producer as well as in consumer behaviour. On the other hand, in an aggregate reduced-form type of model it may be impossible to identify effects both along the dimension of symmetry/asymmetry and with respect to the effects for consumer/producers. For the two oil-exporting countries we believe that it is essential to capture the impacts from the significant structural changes that have taken place in their economies, as they have gone through the transition of becoming net oil exporters. This change of position will in itself imply that asymmetric effects have been operative even if prices have moved in the same direction, and indications of the existence of asymmetry may therefore be obtained from the two oil price variables already included.[8] Clearly, to get good statistical estimates on the two specified price effects requires that the prices are not too strongly correlated.

13.5 EMPIRICAL RESULTS

13.5.1 Correlation between real growth and oil prices only

Table 13.1 summarizes the empirical results for the specifications with oil prices and real GDP growth only. Let us first consider the outcomes for the USA, Canada, West Germany and Japan, i.e. the results in the first four columns in Table 13.1. Remember also that the econometric specification is identical for these four countries. In the table only the sums of individual lag coefficients for the oil price variables are reported. In addition, in the row next to each of these, exclusion F-statistics for the respective group of coefficients are listed with marginal significance levels in parentheses. For three of these countries the calculations carried out confirm the hypothesis that there is a significant negative correlation between increases in oil prices and the subsequent impacts on the overall activity level. The effects are significant at a 5% level for the USA and West Germany and borderline significant for Canada. For the USA, these results are consistent with the previous studies, Hamilton (1983) and Mork (1988).

[8] A special form of 'exogenous' asymmetry may be relevant in this type of model, namely if there are tendencies of differences in lags between crude price movements and changes in product prices when oil prices increase and decrease respectively. Typically, one may expect a rather close correlation (say one to two quarters) between changes in the two variables when crude prices go up. On the other hand, when crude prices turn downward, both private oil companies and governments may act to slow down the fall in product prices.

Empirical results

Moreover, the results from Canada and West Germany roughly confirm those obtained by Burbridge and Harrison (1984). The latter also found reasonable correlation effects for Japan. As seen from Table 13.1, however, we have been unable to detect any significant effect from oil price increases on GDP for this country. In fact, the sign of the estimated sum of coefficients has the opposite sign to what we expected, although the coeffiecients are not significantly different from zero. It is not clear what may explain this deviating relation for the Japanese economy. One possibility is the special social organization of productive activities in this country which may have resulted in an increased competitiveness relative to other Western countries after the two oil crises. Furthermore, the Japanese economy appears to have been able to take advantage of the increased energy scarcity in the 1970s in a much better way than did other OECD countries, such as in production and exportation of fuel-efficient automobiles. The conscious attempts of the Japanese government in 1973–74 to avoid conflicts with the oil-exporting countries may also be noted.

The fourth row in Table 13.1 shows the estimated accumulated effects on the GDP growth rate in the various countries of the decline in real oil prices. These coefficients are positive on average, as expected, for the USA, Canada and West Germany. It is interesting that the coefficients are much smaller in absolute value than the corresponding effects of increased prices. The calculations give some support to the theory of asymmetric effects of oil price fluctuations. However, the estimated effects from price decreases are not significantly different from zero for any country.

For the fourth country, Japan, the calculations yielded a negative coefficient for a decreasing oil price, but again the result is not very clear. In principle, high cost of adjustments may turn the net effect negative. However, as mentioned above, a priori one may rather assume the Japanese economy to be relatively flexible, and thus adjustment costs to be moderate compared to other countries.

The overall small estimated effects of price decreases may be evaluated in light of the hypothesis of asymmetric price effects on economic activity. On the one hand, a fall in oil prices implies improved profits and opportunities to expand for oil-consuming industries. For a net importer of oil, the terms-of-trade effect is also unambiguously positive. However, as discussed above, in the short run, a significant change in relative prices may involve various types of costs of adjustment in the economy. This will work in the opposite direction and partly outweigh the positive stimulus on economic activity. The net effect may be close to zero.

The statistical test of pairwise equality of the coefficients for increases and decreases is rejected on a 5% level for the USA and West Germany, but not for Canada and Japan, as shown by the F-statistics in row 7 of Table 13.1. For the latter countries we also are unable to reject the hypothesis that all the oil-price coefficients are zero (see row 9).

Table 13.1 Univariate results

	USA	Canada	West Germany	Japan	UK	Norway
Oil price increases						
1. Sum of coefficients	−0.10	−0.20	−0.11	0.01		
2. Exclusion test, $F(4, 72)$	3.47	2.41	3.84	0.72		
3. Exclusion test, p-value	0.012	0.057	0.007	0.579		
Oil price decreases						
4. Sum of coefficients	0.004	0.07	0.03	−0.01		
5. Exclusion test, $F(4, 72)$	1.27	0.37	0.62	1.10		
6. Exclusion test, p-value	0.291	0.833	0.647	0.363		
Test of pairwise equality						
7. $F(4, 72)$	1.92	1.09	2.86	1.00		
8. p-value	0.117	0.368	0.030	0.412		
Joint exclusion test						
9. $F(8, 72)$	2.28	1.41	2.10	0.90		
10. p-value	0.031	0.208	0.047	0.523		
Crude price while net importer						
11. Sum of coefficients					−0.02	−0.02
12. Exclusion test, $F(4, 68)$					0.56	0.84
13. Exclusion test, p-value					0.692	0.505
Crude price while net exporter						
14. Sum of coefficients					0.05	0.13
15. Exclusion test, $F(4, 68)$					0.59	2.48
16. Exclusion test, p-value					0.671	0.052
Product prices						
17. Sum of coefficients					−0.14	−0.18
18. Exclusion test, $F(4, 68)$					0.57	3.02
19. Exclusion test, p-value					0.685	0.024

Table 13.1 (Cont.)

	USA	Canada	West Germany	Japan	UK	Norway
Joint exclusion test						
20. $F(12, 68)$					0.71	2.07
21. p-value					0.736	0.031
Standard error of regression	3.84	3.78	4.90	4.59	6.34	11.46

The models for the UK and Norway are specified to capture the effects of the change in economic structure in these countries as they have moved to become net oil exporters. The results for these countries may also shed additional light on the asymmetry hypothesis.

If we focus first on the figures for the UK, we see that the positive effect on GDP from crude oil prices is rather weak and insignificant. One possible explanation for this may be the existence of asymmetric effects, i.e. that structural changes in the British economy have counteracted the positive income effects from increased oil prices. The coefficient for the product price is negative, as expected, but again the significance is poor. In fact, we are unable to reject the hypothesis that all oil price coefficients are zero (see Table 13.1).

For Norway, the picture is the same as for the UK with respect to the sign of the coefficients. However, the positive effect on the GDP growth rate from oil price increases (during the period as an oil exporter) is much stronger – the sum of the coefficients is 0.13 as opposed to 0.05 for the UK. Moreover, for Norway this influence is statistically significant. The impact of product prices is estimated to be more or less the same in the two countries, but again the Norwegian effect is determined more accurately. The higher significance for Norway is remarkable given the much higher residual variance in this equation.

The differences in magnitude and significance between these two countries may be discussed from different points of view. One explanation may be that the relative importance of oil in the economy is much smaller in the UK than in Norway. An equally important factor is probably differences in economic policy pursued in the two countries during much of the observation period. In Norway, economic policy in the latter part of the 1970s was expansionary and consciously directed towards avoiding large increases in unemployment. Some of the expected future oil incomes was spent 'in advance', and while other countries experienced stagnation, this policy obviously worked to keep up growth in the Norwegian GDP. The results from the estimation of multivariate models presented in the section 13.5.2 may help to clarify this question.

13.5.2 Results with all variables included

Table 13.2. summarizes the estimates for the oil price variables when all macroeconomic variables listed above are included in the regression models. The sum of coefficients and exclusion tests for the macroeconomic variables are given in Table 13.3. From the first row of Table 13.2 we first notice that the effects of oil price increases for the four 'oil importing' countries have become more ambiguous. For the USA the estimate from the simple correlation seems robust; an at least borderline significant negative coefficient of around −0.1 is confirmed by the present calculations. Regarding the impact on GDP growth from decreasing oil prices, a somewhat stronger effect is found for the USA compared to the simple oil price–GDP model. This estimate is strikingly different from that of increases, and pairwise equality is rejected unambiguously. Thus, the USA continues to show evidence of asymmetric price effects on the overall activity level.

However, the results for Canada and West Germany are markedly affected by the inclusion of additional variables. For West Germany the GDP effect of increasing oil prices is reduced in absolute terms and has become statistically insignificant. More noticeably, for Canada the sum of coefficients is positive in the multivariate model, although with a low value for the test variable. Explanations for this unexpected result for Canada, which deviates considerably from that for the USA, may be sought along different lines. First, a large domestic production and the strong regulation of petroleum activities is clearly an important part of the picture. Canadian energy policy may have been adjusted to external events in the markets. Second, a high positive coefficient for OECD industrial production is obtained in the multivariate regression model (together with Japan, the highest effect among the countries included). This partly reflects Canada's heavy dependence on the USA as a trading partner.

Altogether, the different findings for Canada can then be interpreted in the following way: oil price increases may benefit (or at least does not hurt) the Canadian economy in a direct sense because of the country's oil industry. Taking account of both policy actions and the response from export markets, however, the total effect may be negative, as indicated by the simple correlations.

For West Germany, the effect of the OECD activity level plays a less important role in explaining the observed GDP fluctuations. Instead, the inflation rate comes out with a rather high negative coefficient. This probably reflects the anti-inflationary policy that has been pursued in West Germany for long periods since the first oil price shock.

Regarding the effects of decreasing oil prices, Canada now shows no sign of asymmetry: in a partial sense, price declines hurt as much as increases are benficial. For West Germany, however, there are only minor changes in coefficient values for the simple correlations. For this country there are still some indications of asymmetric effects (although the test criterion must be relaxed considerably, see row 5 of Table 13.2).

For Japan, even in the univariate model we were unable to detect any significant GDP–energy price correlation. The mulivariate model does not add much to the picture for this country.

Coming then to the oil-exporting countries, we notice first that for the UK, the estimated effects of the oil price variables included still have the expected sign, but the statistical fit is poor. Furthermore, the coefficient for the product price is much smaller than in the univariate case. For Norway, the introduction of more explanatory variables in the model has led to less

Table 13.2 Multivariate results

	USA	Canada	West Germany	Japan	UK	Norway
Oil price increases						
1. Sum of coefficients	−0.10	0.21	−0.05	0.08		
2. Exclusion test, $F(4, 52)$	2.26	1.46	0.88	0.38		
3. Exclusion test, p-value	0.075	0.229	0.485	0.823		
Oil price decreases						
4. Sum of coefficients	0.08	0.18	0.01	−0.04		
5. Exclusion test, $F(4, 52)$	2.27	1.24	0.79	0.59		
6. Exclusion test, p-value	0.074	0.304	0.535	0.672		
Test of pairwise equality						
7. $F(4, 52)$	3.87	0.50	1.44	0.57		
8. p-value	0.008	0.734	0.233	0.689		
Joint exclusion test						
9. $F(8, 52)$	4.81	2.98	1.57	1.12		
10. p-value	0.0002	0.008	0.156	0.367		
Crude price while net importer						
11. Sum of coefficients					0.001	−0.04
12. Exclusion test, $F(4, 48)$					2.14	0.89
13. Exclusion test, p-value					0.090	0.477
Crude price while net exporter						
14. Sum of coefficients					0.05	0.10
15. Exclusion test, $F(4, 48)$					1.78	1.35
16. Exclusion test, p-value					0.148	0.265
Product prices						
17. Sum of coefficients					−0.03	−0.07
18. Exclusion test, $F(4, 48)$					0.03	1.34
19. Exclusion test, p-value					0.998	2.269

Table 13.2 (Cont.)

	USA	Canada	West Germany	Japan	UK	Norway
Joint exclusion test						
20. $F(12, 48)$					1.25	1.22
21. p-value					0.279	0.300
Standard error of regression	3.27	3.22	4.45	4.34	5.88	11.68

Table 13.3 Results for non-oil variables

	USA	Canada	West Germany	Japan	UK[a]	Norway[a]
OECD index of industrial production						
1. Sum of coefficients	−0.24	0.24	0.08	0.50	0.13	−0.48
2. Exclusion test, $F(4, 52)$	0.80	1.23	1.28	2.94	0.25	0.42
3. Exclusion test, p-value	0.529	0.308	0.290	0.029	0.909	0.791
Inflation						
4. Sum of coefficients	−0.32	0.07	−0.26	−0.56	−0.03	0.08
5. Exclusion test, $F(4, 52)$	1.79	0.28	0.91	2.26	2.63	0.80
6. Exclusion test, p-value	0.145	0.888	0.467	0.075	0.046	0.533
Interest rate						
7. Sum of coefficients	−0.76	−1.30	−0.87	0.02	−0.89	−1.33
8. Exclusion tests, $F(4, 52)$	3.54	7.56	0.94	0.33	0.97	1.17
9. Exclusion tests, p-value	0.013	0.000	0.450	0.855	0.432	0.336
Unemployment						
10. Sum of coefficients	0.87	1.13	−0.58	−4.24	0.49	2.69
11. Exclusion tests, $F(4,52)$	4.96	4.45	1.46	0.84	1.70	0.64
12. Exclusion tests, p-value	0.002	0.004	0.227	0.508	0.166	0.634
Real wage changes						
13. Sum of coefficients	−0.23	0.33	−0.09	0.06	0.02	0.46
14. Exclusion test, $F(4,52)$	0.20	2.35	0.89	0.59	1.81	1.08
15. Exclusion test, p-value	0.939	0.066	0.477	0.674	0.143	0.379

[a] For the UK and Norway, the degrees of freedom for the F-statistics are 4 and 48.

significant energy price effects. Both the specified price coefficients have retained their signs and magnitudes from the simple correlation model, although both have become insignificant, especially the positive income effect of the crude price. The explanation for this outcome may, of course, be simply the loss of degrees of freedom in the regression. However, one may also stress a couple of interesting features from the estimated coefficient values of the macroeconomic variables. For the OECD production indicator, a negative value is obtained (see Table 13.3). This may be interpreted as an indication of the contra-cyclical economic policy that was conducted during the 1970s and early 1980s in Norway. The expansionary policy measures taken in this period may also provide a partial explanation for the strong positive correlation between GDP and the unemployment rate for Norway.

13.6 SUMMARY AND CONCLUSIONS

The purpose of this chapter has been to investigate the relation between oil price fluctuations and business cycles. A particular emphasis has been on the hypothesis of asymmetric effects on GDP of changes in oil prices. Possible theoretical explanations for asymmetry have been discussed. An aggregate, reduced-form type of model has been specified and estimated for a number of countries which differ, in particular, with respect to the net trade position. Consistent with the findings of previous authors in this field, the empirical results show significant correlations between oil prices and GDP growth for the USA. For this region, we also find strong support for asymmetric price effects. For the other countries included in the study, the conclusions are not as sharp as for the USA. When linking changes in GDP growth to fluctuations in oil prices only, significant correlations are found for Canada and West Germany. However, in the more elaborate model also including a set of other macroeconomic variables, the significance of the oil price variables more or less disappears. This should not be too surprising. One problem with the kind of analysis we have carried out is of course the data used in the estimations. As mentioned above, specific efforts have been undertaken to construct a relevant oil price variable for the USA. Moreover, within the multivariate model there may be problems both with degrees of freedom and multicollinearities between the explanatory variables. In particular, we have pointed out that both regulations of domestic energy markets (Canada) and specific economic policy measures triggered by events in the oil market may affect economic activity via some of the specified variables. The study also include calculations for two countries, the UK and Norway, that have moved into a net export position for oil during the observation period. For these countries, one should expect opposite effects on GDP growth when oil prices change compared to those of an oil-importing country. However, structural changes and adjustment costs may dampen the stimulus from, for example, increasing prices on petroleum. This may be part of the explanation why the estimated effects from oil prices on growth in the UK are so weak. For Norway, on the other hand, strong correlations are obtained in the simple correlation model. The size of the oil sector and the expansionary economic policy pursued in the 1970s are probably important underlying factors. As for most other countries, the direct effects from oil price fluctuations become less significant when other macoeconomic variables are included.

APPENDIX

Variables and data sources

Except for the unemployment rate and the interest rate, all variables have been included in the regression with their yearly growth rates, calculated as 400 times the log change of the respective quarterly figures. The length of the available time series varied between variables and from country to country. We decided to use the same time period for all countries, 1967:2–1988:2.

Real GNP/GDP

For the USA, Canada, Japan and West Germany we have used the GNP in constant prices (seasonally adjusted). The source here has been OECD MEI. For the UK and Norway we have used data on GDP (in constant prices and seasonally adjusted) available in MEI and from the quarterly National Accounts from the Central Bureau of Statistics, Norway respectively.

Interest rate

For the USA, Canada and the UK we have used the 'treasury bill rate', for West Germany 'rates on 3-month loans, Frankfurt' and for Japan 'call money rate', all available in MEI. For Norway we have used a long-term interest rate (yield on government bonds (−1985:3)) available in MEI and 'effektiv rente på statsobligasjoner' from the quarterly journal 'Penger og kreditt' issued by the Bank of Norway.

Real wages

For all countries except the UK and Japan, we have used 'hourly earnings'. For the UK we have used 'weekly earnings' and for Japan 'monthly earnings'. All data series are seasonally adjusted and available in MEI.

Unemployment rate

For all countries except Norway we have used figures for 'unemployment as per cent of total labour force' in MEI, seasonally adjusted. For Norway we have used an equivalent measure from a quarterly labour force sample survey (AKU, Central Bureau of Statistics).

Inflation rate

The inflation variable was constructed by dividing GNP/GDP at current prices by GNP/GDP at constant prices (GNP/GDP deflator). The source for all countries except Norway is MEI. For Norway the data source is the quarterly National Accounts from the Central Bureau of Statistics.

Overall index of industrial production in the OECD-area

The same index for total OECD production available in MEI has been used for each country.

Oil price variables

For the USA, we have used a PPI for crude oil corrected for price controls constructed by Mork (1988). This index represents a linking of the PPI for crude oil with the refiner acquisition cost (composite for imported and domestic oil).

For Canada, West Germany and Japan we have used the PPI for petroleum products (for Japan petroleum and coal) available in MEI.

For Norway and the UK, we have used the PPI for petroleum products. The Norwegian index is available in MEI, while the UK index represents a linking of the PPI for petroleum and coal products until 1974 (MEI) with the PPI for petroleum products from the Business Statistics Office (Department of Trade and Industry). In addition, we have used the world price of crude oil (Arabian light) measured in local currencies. Unil 1978:4 'posted prices' have been used (OPEC publication: *Annual Statistical Bulletin*). Thereafter 'spot prices' (*OPEC Bulletin*) have been applied.

All oil price variables have been deflated with the PPI (MEI) for the various countries.

REFERENCES

Berger, K., Cappelen, Å., Knudsen, V. and Roland, K. (1987) The effects of a fall in the price of oil: the case of a small oil exporting country, in *Economic Modelling in the OECD Countries* (ed. H. Motamen), Chapman and Hall, London.

Burbridge, J. and Harrison, A. (1984) Testing for the effects of oil price rises using vector autoregressions. *International Economic Review*, **25**(2), 459–84.

Darby, M. R. (1982) The price of oil and world inflation and recession. *American Economic Review*, **72**(4), 738–51.

Davis, S. J. (1987) Allocative disturbances and specific capital in real business cycle theories. *American Economic Review, Papers and Proceedings*, **77**, 326–32.

Gisser, M. and Goodwin, T. H. (1986) Crude oil and the macroeconomy: tests of some popular notions. *Journal of Money, Credit, and Banking*, **18**(1), 95–103.

Gordon, R. J. (1975) Alternative responses to external supply shocks. *Brookings Paper of Economic Activity*, **1**, 183–204.

Hamilton, J. D. (1983) Oil and the macroeconomy since World War II. *Journal of Political Economy*, **91**(2), 228–48.

Hamilton, J. D. (1988) A neoclassical model of unemployment and the business cycle. *Journal of Political Economy*, **96**, 593–617.

Horwich, G. and Weimer, D. L. (1984) *Oil Price Shocks, Market Response and Contingency Planning*, American Enterprise Institute, Washington D.C.

Huntington, H. G. (1984) Real oil prices from 1980 to 1982. *The Energy Journal*, **5**, 119–31.

Longva, S., Olsen, Ø. and Strøm, S. (1988) Total elasticities of energy demand analyzed within a general equilibrium model. *Energy Economics*, **20**, 298–308.

Loungani, P. (1986) Oil price shocks and the dispersion hypothesis. *Review of Economics and Statistics*, **58**, 536–9.

Mork, K. A. (1988) Oil and the macroeconomy when prices go up and down: an extension of Hamilon's results, manuscript, Vanderbilt University, September. Forthcoming in *Journal of Political Economy*.

Mork, K. A. and Hall, R. (1980) Energy, prices, inflation and recession, 1974–1975. *The Energy Journal*, **1**, 31–63.

Phelps, E. S. (1978) Commodity-supply shock and full-employment monetary policy. *Journal of Money, Credit, and Banking*, **10**(2), 206–21.

Pierce, J. L. and Enzler, J. J. (1974) The effects of external inflationary shocks. *Brookings Paper of Economic Activity*, **1**, 13–54.

Sims, C. A. (1980) Macroeconomic and reality. *Econometrica*, **48**, 1–47.

van Wijnbergen, S. (1985) Oil discoveries, intertemporal adjustment and public policy, in *Macroeconomic Prospects for a Small Oil Exporting Country* (eds O. Bjerkholt and E. Offerdal), Martinus Nijhoff, Dordrecht.

Author index

Aaheim, A. 103–23, 144, 160
Aarrestad, J. 103
Abodunde, T. T. 207
Adelman, M. A. 3, 9, 175
Alcamo, J. 89
Alfsen, K. 87–100
Anand, S. 175
Arrow, K. J. 173, 188
Aslaksen, I. 103–23, 144, 182, 187–205

Bailey, M. F. 175
Bartlett, S. 29–47
Berger, K. 221–38, 243
Bergmann, B. 14
Bertrand, J. 23
Binmore, K. 56
Bjerkholt, O. 3–28, 68, 109n, 115, 182, 187–205, 221–38
Blitzer, C. R. 3
Brekke, K. A. 22, 103–23, 144, 187–205
Brennan, M. J. 189
Brooke, A. 158
Burbridge, J. 239, 246, 249
Byatt, I, 128
Bøhren, Ø. 176

Cappelen, Å. 125–52, 243
Christophersen, Y. 175
Corden, W. M. 125
Cox, L. C. 3

Dagsvik, J. 30, 37
Dahl, C. A. 9
Darby,, M. R. 239
Dasgupta, P. S. 102, 223
Davis, S. J. 240, 242, 247
Debreu, G. 174

Dixit, A. 78, 82

Eastwood, P. K. 103
Ekern, S. 176, 188
Eliassen, A. 90
Enzler, J. J. 239

Farzin, Y. H. 164n
Fisher, A. C. 188
Friedman, J. W. 71
Førsund, F. 154

Gately, D. 223
Gisser, M. 239
Gjelsvik, E. 3–28, 38, 43, 68, 106, 125–52
Goett, A. 35
Golombek, R. 8, 76, 153–69
Goodwin, T. H. 239
Gordon, R. J. 239
Goudswaard, K. 128
Grinols, E. L. 174
Grossman, S. J. 78, 177
Grout, P. A. 209

Hall, R. 239
Hamilton, J. D. 239–44, 248
Hanemann, M. 188
Harrison, A. 239, 246, 249
Hart, O. D. 78, 208
Hawk, D. 38n
Heal, G. 103, 223
Hoel, M. 8, 49–65, 68, 69, 153–69, 208, 214, 223, 228
Holden, S, 228
Holtsmark, B. 49–65
Hopper, R. J. 10n
Hordijk, L. 89
Horwich, G. 241

Author index

Hotelling, H. 223
Huntington, H. G. 246n
Hurst, C. 17

Jensen, M. C. 175
Johansen, L. 173
Johnsen, T. A. 103–23, 144
Johnson, H. G. 153

Kämäri, J. 89
Kar, H. van de 128
Kartevoll, T. 173, 179
Kauppi. P. 89
Kendrick, D. 158
Ketoff, A. 38n
Klein, L. R. 93
Kline, D. M. 223
Knudsen, V. 243
Kobila, T. Ø. 187–205

Lemhaus, J. 90
Levine, M. D. 35
Lind, R. C. 173, 175
Lindstrøm, T. 187–205
Lintner, J. 174
Lommerud, K. E. 174
Longva, S. 243, 245
Lorentsen, L. 30, 87–100, 173, 179, 223
Loungani, P. 240, 247
Lund, D. 171–86
Lynch, M. C. 3, 9

Maddala, G. S. 38
Majd, S. 188
Mankiw, N. G. 181
Manne, A. 58n
Massé, P. 187
McDonald, R. 188, 191, 203
McFadden, D. 35
McMahon, J. E. 35
Meeraus, A. 158
Messner, S. 9
Meyers, S. 38n
Moore, J. 208
Mork, K. A. 239–60
Mossin, J. 174
Mysen, H. T. 239–60

Nalebuff, B. 175
Newbery, D. 214, 223
Nyborg, K. 87–100

Odell, P. 3, 5, 14, 21, 88
Offerdal, E. 136–42
Øksendal, B. 187–205
Olsen, T. E. 188
Olsen, Ø, 3–47, 30, 38, 43, 68, 106, 221–60

Paddock, J. L. 188
Parsons, J. 3
Phelps, E. S. 239
Pierce, J. L. 239
Pindyck, R. S. 182, 188, 189, 207
Posch, M. 89
Roberts, K. 188
Rogner, H. 21
Roland, K. 58n, 223, 243
Rosland, A. 94
Rubinstein, A. 56
Ruderman, H. 35
Runca, E. 89
Russell, J. 158

Saltbones, J. 90
Sandmo, A. 174, 176
Schipper, L. 38n
Schmalensee, R. 174
Schwartz, E. S. 189
Shapiro, M. D. 181
Sharpe, W. F. 174
Shiller, R. J. 177
Shiryayev, A. N. 188
Siegel, D. 188, 191, 203
Sims, C. A. 240, 244
Skånland, H. 104
Smith, J. L. 188
Stauffer, T. R. 104
Steigum, E. 144
Stensland, G. 188
Stephen, G. 58n
Stern, N. 82
Strubegger, M, 9
Strøm, S. 29–47, 58n, 68, 104, 136–42, 173, 179, 243, 245
Sutton, J. 56

Tiago de Oliveira, J. 37n

Vatne, B. H. 22, 38, 43
Vislie, J. 49–85, 154, 207–17

Ward, M. 104

Weimer, D. L. 241
Weitzman, M. L. 188
Weyant, J. P. 223
White, D. C. 3

Wijnbergen, S. van 125, 127, 243
Wirl, F. 207, 216–17
Wolinsky, A. 56
Wright, A. 3

Subject index

Acid rain 88
Algeria 24, 57, 158
   natural gas
      reserves 7
      production 7
   future gas supply in deregulated EC market 25
AMEN model 136
Asset pricing 174
Asymmetry
   in price responses 241–44
Autocorrelation 37

Bargaining 50, 68
   game 58
   outcome 74
   theory 51
   position 55
Bertrand, *see* Games
Beta 176
Bilateral
   monopoly 49, 69, 208
   oligopoly 49, 67, 207
Booming sector 126
British Gas 49, 158
   gas price policy 18
Brownian motion 188, 190
Business cycles 239–58

Canada 248
Capacity utilization 37
CAPM (Capital asset pricing model) 174, 191
Certainty equivalent 118
Claim
   contingent 172
   non-contingent 172
Collusion 159
Common carrier principle 13–15, 78
Conjectural variation 158

Consumer surplus 10–13
Contract
   long-term 62, 158, 208, 215
   rules 72
   self-enforcing 208, 214
   complete 208
   natural gas 158
   take-or-pay 15
Core
   of a game 53–4
   of the European gas market 56
Cost
   structure 57
   extraction 57
   marginal 12, 156
   long-run marginal 157
   hydroelectric power 154
   gas-fired thermal power 154
   transportation 57, 154
   environmental 164
Hydroelectric
   power capacity 154
   power 154
Cost–benefit analysis 179
Cournot 159, *see also* Games
Covariance
   measure 176

Demand
   electricity 160
   elasticity 158
   curve 154
Deposition
   of sulphur 93
Deregulation 3, 67, 77
Dioxide
   carbon 87–99, 164
   nitrogen, 164
   sulphur, 87–99
Disagreement point 55, 211

Discount rate 191
  social 171–82
Discrete–continuous choice 30
Distribution company 154
DOM (Dynamic oligopoly model) 22–7
Dutch disease 125–7, 243
  adverse effects on manufacturing
Dynamic (time) consistency 214
Dynamic programming 23

EC 1992 13
EC Commission 77
  energy policy 15
EMEP (Programme for monitoring and evaluation of the long range transmission of air pollutants in Europe) 90
Emissions to air
  see Dioxide
  forecasts 97
Energy
  price elasticity 38
Envelope theorem 184
Environmental problems 88–99
Expectations
  static 116
  dynamic 117
Expected return 109

Financial asset 184
Fluctuations
  in economic activity 239
  in oil prices 240–1
France 17, 57
  natural gas
    consumption 6
    production 7
    reserves 7
    residential consumption 31
  netback 19
  effects of unit cost pricing 21
Fuel
  switching 30, 87
  choice 35–6
  transition 36, 88–9, 94
  shares, projected 41

Games
  Bertrand price 23
  Cournot investment 23
GEM model 20
Greenhouse effect 99

Hamilton-Jacobi-Bellman equation 196–7
Hamiltonian equation 166
Hotelling
  rent 153

IEA 58, 70
Import diversification 57
Individual rationality 212
International jurisdiction 208
Intertemporal
  utility 114
  optimization 114
Irreversibility 189–94
  premium 193
Italy 57
  natural gas
    consumption 6
    production 7
    reserves 7
    residential consumption 31

Japan 248
Joint production 208, 215

KVARTS model 126

Lawrence Berkeley Laboratory 38
LINK economic modelling project 93

Market power 154
Market diversification 174
Market share requirement 58, 70
Market value 174
Markov process 37
MNL (multi-nomial logit choice) 35
MODAG model 126

Nash
  bargaining solution, 55, 71, 211
  equilibrium 23, 80
  solution 23
National wealth 103–23
  optimal management 114–22
  calculation of 105–7
Natural gas supply regions
  Groningen 7
  Algerian Sahara 7
  Uringoi (Sibiria) 7
  North Sea 7
Natural gas reserves 7
Natural gas market
  Western Europe 3–27

deregulation 77
distribution network
  structure 71
Natural gas costs
  in production 9
  in transmission 9
  in distribution 9
Natural gas demand 29
  residential 30, 39
  in Western Europe 30
  for space heating 33
  projections 39
Negotiations 51, 72, 207
Net present value 235
Netback 18–19
Netherlands 57, 158
  natural gas
    consumption 6
    production 7
    reserves 7
    residential consumption 31
Non-OPEC countries 221
Non-traded asset 176
Norway 24, 56, 69, 171, 154, 250
  natural gas
    production 7
    reserves 7
    unit costs 58
    price range projections 59
    export projections 61
  future gas supply in deregulated EC
    market 25

Oil futures contract 174
Oil fund 145
OPEC 221–38
  breakdown 229–31
  cooperation 231–4
Optimal extraction strategy 172
Optimal stopping 203
Option value 189
Outside option principle 56

Perestroika 76
Permanent income 148–50
  life-cycle theory
Petroleum wealth 105, 149
  definition of 107
Petroleum tax 175
Petroleum-exporting country 171
Power generation
  hydro 187
  thermal 187

Preference function 115–19
  quadratic 115
  exponential 118
Present value
  risk-adjusted 118
Project evaluation 180
Purvin and Gertz
  gas demand forecasts 24

RAINS (Regional acidification
  information and simulation) model
  89
Rate of interest 163
Rational expectations equilibrium 215
Rate of return
  required 192
Rate of pure time preference 172
Rent
  monopoly rent 12
  resource rent 12, 153
  rent from price discrimination 12
Reservation price 50, 190, 200
Resource
  rent 156
Risk
  systematic 176
  aversion 181
  measure 176
    adjusted discount rate 174
Roy's identity 37
Ruhrgas 49, 158

Scarcity value of oil 213
Shadow price
  of oil 171–3
SNA (Standard of national accounts)
  103
Social discount rate 171
Statistical test 249
Stochastic process 190
Stochastic control 195
Stochastic dynamic programming 183
Stock market 174
Strategy 183
  flexible 106
  constant 106
Subjective probability 106, 172

Tax wedge 173
Credit rationing 180

Transmission companies 10
Transportation matrix 90

Uncertainty
  geological 179
  technological 179
United Kingdom 17, 56, 175, 250–2
  natural gas
    consumption 6
    production 7
    reserves 7
    residential consumption 31
    import projections 59
  netback 19
USA 248
USSR 24, 56, 69, 158
  natural gas
    production 7
    reserves 7
    unit costs 58
    price range projections 59
    export projections 61
  future gas supplies in deregulated EC market 24

Value function 183
VAR (Vector autoregressive) model 244

Vertical control 77, 78
Vertical integration 78
von Neumann-Morgenstern utility 171

West Germany 17, 57, 158, 248
  natural gas
    production 7
    reserves 7
    residential consumption 31
  netback 19
  gas price policy 18
  effects of unit cost pricing 21
Western meteorological synthesizing center 90
Western Europe
  natural gas
    consumption 6
    market 51
    production 7
    reserves 7
    import projections 59
    demand projections 44
Windfall profits 126
WOM model 223–9
Wood and MacKenzie 111
World commission on environment and development 87